D0754377

Seismic Exploration for Sandstone Reservoirs

Nigel A. Anstey

SEISMIC EXPLORATION FOR SANDSTONE RESERVOIRS

NIGEL A. ANSTEY

INTERNATIONAL HUMAN RESOURCES DEVELOPMENT CORPORATION
Boston

ISBN: 0-934634-04-1

Library of Congress Catalog Card Number: 80-82135

Printed in the United States of America

Contents

Preface

This text was originally written for use with the videotape program of the same title. Numbered video cassettes correspond to the following chapters of the book:

Tape	Chapter(s)	Tape	Chapter(s)
1	1-2.2.1	11	3.4
2	2.2.2-2.3	12	3.5.1-3.5.2
3	2.4	13	3.5.3
4	2.5-2.6	14	3.5.4-3.6.3
5	2.7-2.8	15	3.7-3.7.2
6	3	16	3.8-4
7	3.1	17	4.1
8	3.2-3.2.1	18	4.2-4.4
9	3.2.2-3.2.3	19	5
10	3.3		

Complete information about the videotape program, *Seismic Exploration for Sandstone Reservoirs* may be obtained from: IHRDC, 137 Newbury St., Boston, MA 02116, (617) 536-0202.

Acknowledgements

Thanks are expressed to the authors, companies and professional associations who have allowed the use of their material in the course. The author and publisher are grateful for permission to reproduce material whose copyright belongs as follows:

Figures:

2.4–3 (Widess) SEG; 2.4–4 (Prescott) Continental Oil Company; 3–1 (Le Blanc) AAPG; 3–2 (MacKenzie) AAPG; 3.1–1 Seiscom Delta; 3.1–3 (Schramm et al.) AAPG; 3.1–4 (Larner et al.) Western Geophysical Company; 3.2.2–2 Prakla Seismos; 3.2.2–3 (Leung et al.) Amoco Europe; 3.2.2–4 (Newman et al.) S&A Geophysical; 3.2.3–2 Seiscom Delta; 3.3–1 (Taner) Seiscom Delta; 3.4–1 (Weber et al.) Applied Science Publishers; 3.4–2 to –7 (Bruce) AAPG; 3.4–8 Seiscom Delta; 3.5.2–1, –2 (Selley) Academic Press; 3.5.2–3 (Eckelmann et al.) Applied Science Publishers; 3.5.3–2 (Tufekcic) EAEG; 3.5.3–3 (Lyons et al.) AAPG; 3.5.3–5 (Farr) Western Geophysical Company; 3.5.4–1 (Le Blanc) AAPG; 3.5.4–2, –3 (Busch) AAPG; 3.5.4–4, –5 (Lindsey, et al.) World Oil; 3.6–1, –2 (Mitchum et al.) AAPG; 3.6–3 (Busch) AAPG; 3.6–4 (Gould) SEPM; 3.6–5 Prakla Seismos; 3.6–6 Aquatronics; 3.7–1, –2, –6 (Selley) Academic Press; 3.7–4 (Schramm et al.) AAPG; 3.7–5 Prakla Seismos; 3.7–7 Seiscom Delta; 3.7–8 Prakla Seismos; 3.8–1 (Moore) GSA; 3.8–2 (Selley) Academic Press; 3.8–3 (Brown et al.) AAPG; 3.8–4 Seiscom Delta; 3.8–5 (Walker) AAPG.

Film material:

This Land, Shell Oil Company
Book Cliffs Field Trip, part 1, Shell Oil Company
The Beach: A River of Sand, Encyclopedia Britannica

Slides:

17a, 120 (Selley) Academic Press; 18-22 (Scholle) AAPG; 42, 47 (Neidell et al.) AAPG; 43, 46, 70-74, 78, 11A, 114, 125, 154A, 161, 162, 168, 169, 180, 234, 236, 248-252, 275, 276, 288-9, 294-6, 305-6, 308-9, 311-5, 317-8 Seiscom Delta; 44, 61 (Larner et al.) Western Geophysical Company; 51B (Kendall) SEG; 63, 90, 112, 115, 116, 136, 231, 232, 240, 267, 290-1, 303-4, 307, 321 Prakla Seismos; 75, 111, 133, 137 (Mitchum et al.) AAPG; 76, 319 S&A Geophysical; 88 (Le Blanc) AAPG; 94, 95, 101 (Barry et al.) Teledyne Exploration; 96 (Patch) Geocom; 96 (Backus et al.) EAEG; 102, 104, 285-7 (Schramm et al.) AAPG; 113 (Leung et al.) Amoco Europe; 123, 124, 131, 259, 263, 265 (Sangree et al.) AAPG; 127, 228, 281 (Coleman) Continuing Education Publication Company; 134, 220 United Geophysical Corporation; 139, 273-4 (Vail et al.) AAPG; 146 USGS; 158, 316 Western Geophysical Company; 159, 160 Chevron; 163 (Lund et al.) Oil & Gas Journal; 167 (Tucker) AAPG; 177 (Spearing) GSA; 197, 260 (Stuart et al.) AAPG; 208 (Exum et al.) AAPG; 216A, 245 (McGregor et al.) AAPG; 258 (Mitchum et al.) AAPG; 261, 266 USGS; 277-9 (Busch) AAPG; 280 (Fisher et al.) AAPG; 292-3, 322 (Lavergne et al.) Institut Francais du Petrole; 301 (MacKenzie) AAPG.

1 Introduction

"Seismic?" he said, "seismic, for finding strat traps? No way. I'll tell you what you need to find strat traps—courage, and faith, and optimism, and the free-enterprise system, and the money to drill 34 dry holes."

When he said it, it was fair comment. Geophysicists had been talking for several decades about seismic detection of non-structural sandstone reservoirs, but it was mostly just talk.

Then came the bright-spot techniques, and conspicuous success. But the success was limited to gas, and to favourable types of reservoir; it remained true that—except in bright-spot country—not many crews were out there specifically looking for stratigraphic traps.

That is not to say that the **hope** did not persist. Many times we had all seen stratigraphic sandstone reservoirs on seismic sections—as minor anomalies of reflection character—**after** we knew they were there. But seismic sections are full of minor anomalies, and drilling every such occurrence would be out of the question. What we needed was an assurance that a particular anomaly was real, and an interpretive regime which allowed us to select those anomalies likely to represent reservoir situations.

No one would pretend that in 1979 we are there. But we are making progress. As usual, the progress has not come by a single breakthrough, but by a number of steady advances in different parts of our science. Specifically, these are:

- a continuing improvement in seismic recording and processing, to the point where we begin to **trust** quite minor subtleties in the wiggles,
- a continuing improvement in seismic display, to the point where we no longer miss **seeing** these subtleties,
- an improved understanding of the depositional environments in which sandstone reservoirs originate,
- the techniques of seismic stratigraphy, which formalize the relationships between a succession of depositional environments and their seismic expression, and
- last but far from least, the improved dialogue and understanding between the geologist and the geophysicist.

This seems a good time to summarize where we are along the stream of these developments; that will be our task in this course.

What we shall do is this.

First we shall review briefly what it is about a geological feature which makes it visible or invisible on a seismic section.

Then we shall apply those conclusions to many of the types of sandstone reservoir identified for us by the geologists.

Where we decide that a target is likely to be visible, we shall use the seismic method to search for it directly. If the target is likely to be visible only when we know it is there (for example, after a discovery), we shall use the seismic method to locate step-out wells.

Where we decide that a target is **not** likely to be directly visible, we shall use the seismic method to search for the **environment** where sandstone reservoirs are geologically probable.

Then we shall discuss what can be done to improve the seismic indications, as soon as the nature of a target is known.

And finally we shall talk about the contribution of the borehole geophone to the delineation of a reservoir.

The course, then, is a synthesis of seismic technology, of seismic stratigraphy and of sandstone geology. For the first, repeated reference will be made to the author's *Seismic Interpretation: The Physical Aspects* (abbreviated SITPA). For the second, repeated reference will be made to the classic AAPG Memoir 26, *Seismic Stratigraphy* (abbreviated SS). For the sandstone geology, repeated reference will be made to AAPG Memoir 21, *Stratigraphic Traps in Sandstones: Exploration Techniques* (Busch) and AAPG Memoir 16, *Stratigraphic Oil and Gas Fields: Classification, Exploration Methods, and Case Histories.* Sandstone geology is also expounded in a number of current courses by Busch, Klein, Berg, Selley and others; the Klein course is available on videotape.

2 Can We Hope To See It?

In this section we ask what it is about a geologic feature which decides whether it is visible or invisible to the seismic method. We shall do this with a minimum of discussion; participants interested in a fuller account should consult the given page references in SITPA. Many references to original papers are also given in SITPA.

2.1 Will it give a reflection?

The property of the geology which determines the generation of reflections is **acoustic impedance**; loosely, we may think of acoustic impedance as acoustic **hardness**. This is not quite hardness in the sense of the geologist's scale of **scratching** hardness; it is more like hardness in the sense of the man-in-the-street—how difficult the rock is to **squeeze** (SITPA 45-47).

A reflection is generated at the interface between two materials of different acoustic impedance (SITPA 101-110). The more different the hardness of the two materials, the stronger the reflection.

If the first material is softer than the second, the reflection is positive; if the first material is harder, the reflection is negative. If a positive reflection on a section has a certain pulse shape, then a negative reflection has the same shape but upside-down—every peak a trough, every trough a peak.

The **contrast** between two geological materials therefore defines the **strength** and the **polarity** of the reflection from the interface between them.

There is no such thing as a reflection from such-and-such a formation—only from the contrast between that formation and the one above it.

Many or most seismic reflections correspond to time-stratigraphic horizons or bedding planes (SS 51; Sangree and Widmier, 1979). We shall find that there are significant and important exceptions to this, but we shall accept the statement as a generalization for the moment.

It is very important to note that a time-stratigraphic horizon of great geological significance may not generate a visible reflection; a notional reflection is present, but if by chance the acoustic impedances above and below the horizon are substantially equal, the reflection has substantially zero amplitude.

Thus if we visualize near-shore sedimentation embodying a change of facies, and we hypothesize a transgression so rapid that the contact is effectively a time-stratigraphic horizon, then that contact may be visible as a strong reflection in some places—but quite invisible in others. It all depends on the local contrast of acoustic hardness between the sediments deposited before and after transgression (figure 2.1-1).

Just as there is no guarantee that a **time-stratigraphic** boundary must give a visible reflection, so there is no guarantee that a **lithologic** boundary must give one either. Often, it is true, different lithologies have different acoustic impedances, and so generate a seismic reflection; but there is no guarantee of this. The reason

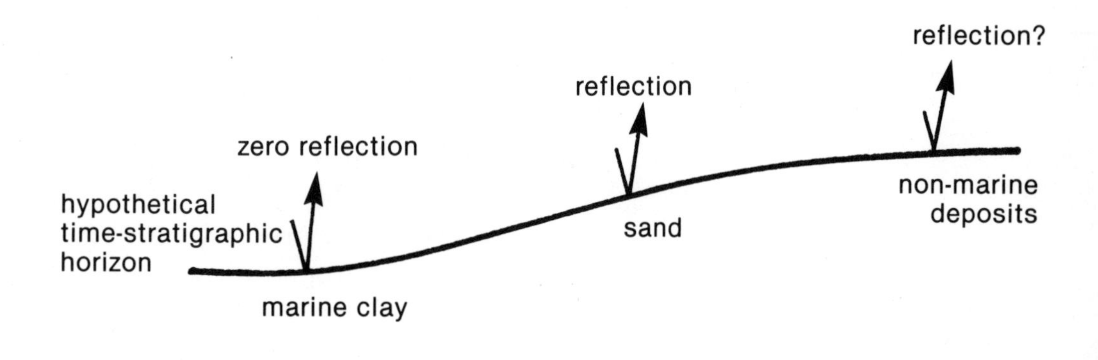

FIGURE 2.1-1

is that the acoustic impedance of a certain lithologic sample varies with many factors: its depth, its tectonic compression, its burial history, its intergranular porosity, its fracture porosity, its fracture type, its cementation, and its saturant (figure 2.1–2).

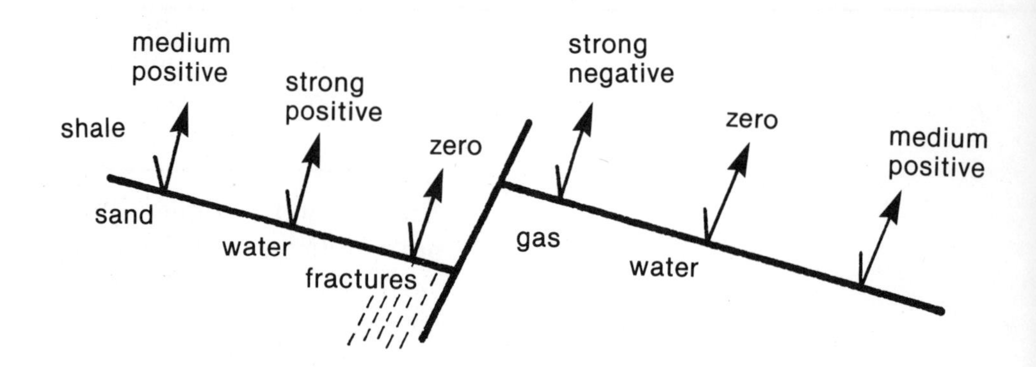

FIGURE 2.1–2

In particular, the boundary between a shale and a sandstone may represent either an increase or a decrease of acoustic impedance, depending on these factors; thus it may generate a positive reflection, or a negative reflection, or none at all.

Even in less extreme cases, we have all seen sections where reflections just come and go—where there **must** be continuous time-stratigraphic horizons, but where there are no continuous reflections.

In the present context, obviously, this is a major problem. So we should look at it in a little more detail.

2.2 Summary of the important properties of rocks (SITPA 51-99)

In a seismic sense the important properties of rocks are their velocities (which decide the seismic ray-paths) and their acoustic impedances (which decide the strength of reflections).

The acoustic impedance or acoustic hardness is the product of the velocity and the density. Density is thus an important property also. Therefore there are three important properties: velocity, density and acoustic impedance—any two of which define the third.

There is a suggestion that absorption is a useful property for distinguishing between rock types; however, there are major problems in measuring it in practice (SITPA 118-134, 142, 213-219, 230, 251, 252, 277-284, 299, 318, 333, 389, 390). Here we shall concentrate on velocity, density, and their product the acoustic impedance.

2.2.1 Shales

An important characteristic of shales is their readiness to compact after deposition. In the geological sense, this accounts for the frequent observation of draping of shale sediments over sandstone and carbonate bodies. In the seismic sense, it accounts for a marked dependence of both velocity and density on depth.

Of course, we could equally well say that the dependence is on **porosity**. At the time of deposition, a fine-grained clay material may have a water content of 60% or more, a relative density of 1.7, and a velocity of perhaps 1600 m/s (5250 ft/s); after extended deep burial and the progressive squeezing-out of water back to the surface, it may have a water-filled porosity of only a few per cent, a density of 2.6, and a velocity of 3700 m/s (12000 ft/s). These are very large ranges—a range of 1.5 to 1 for density, 2.3 to 1 for velocity, and 3.5 to 1 for acoustic impedance. The mechanism for this enormous variation is **compaction**.

Figure 2.2.1-1 suggests bands which velocity and density are likely to occupy in shales. The vertical scale at the left is depth; the velocity scale is at the bottom, the density scale at the top. The density band implies a corresponding spread of porosities (more properly water-contents, in this case); the percentage figures are suggested at the right. The change of density and porosity is concentrated at shallow depths; velocity continues to change, with the **ratio** of porosity change, even when the absolute porosity change is small. The small amount of water left in the pores has only a minor effect on density, but a significant effect on velocity.

Therefore depth is the most important factor affecting the seismic properties of shales.

Where two shales at the same depth have different properties, this is usually because:

- one of them has previously been buried to greater depth, or

- one of them has been subject to tectonic compaction in addition to overburden compaction, or
- one has significant carbonate contamination, or
- the rates of burial have been significantly different, or
- one has been at least partially metamorphosed, or
- one has a high content of organic material or gas, or
- one has become sufficiently brittle to fracture.

The fourth of these—different rates of burial—highlights the difference between two situations: the "normal" situation where each increment of overburden pressure produces the appropriate squeezing-out of pore water before the next increment of pressure is applied, and the "abnormal" situation where the water cannot get out in time. The latter situation, which is by no means truly abnormal, represents a degree of over-

FIGURE 2.2.1–1

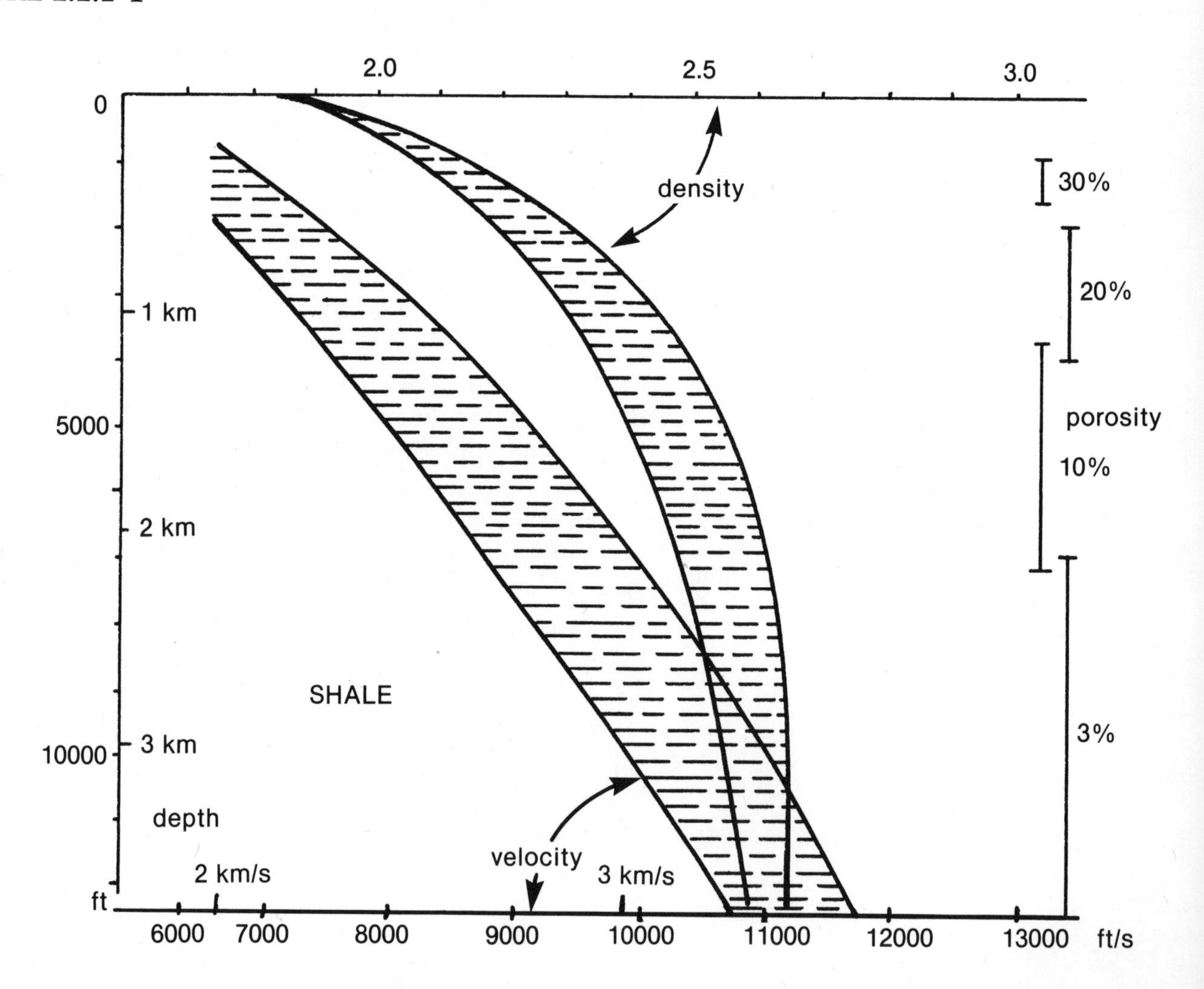

pressure (or undercompaction) in the shale; values of relative density and velocity are much depressed—in extreme cases as low as 1.8 and 2150 m/s (7000 ft/s) (SITPA 57-62, 69, 72, 579-587).

2.2.2 Sandstones

Some geologists use the term "sand" with no connotation other than that of particle size. For seismic purposes, we probably need a distinction of chemical constitution also; let us restrict our simplified discussion to siliciclastic sands.

The simplest case, to consider first, is that of well-sorted beach sands. The sand grains, having been pounded by waves for centuries, are smooth; pounded says rounded. The packing of grains is efficient, approximating to the ideal hexagonal packing of spheres. The pressures of deep burial cannot achieve much compaction by rearrangement of the grains, neither can there be much compaction by squeezing; the contacts are flattish, and the grains themselves are hard. Such sands, therefore, show only a minor dependence of density and velocity on depth. The same is true of mature dune sands.

Rather different, however, is the case of sands in a stream-cut channel. These are likely to be transported only a short distance, and subjected to rapid burial without reworking. The grains are poorly-sorted and angular, and there is much less guarantee of efficient packing during deposition. When such a sand is subjected to the pressures of deep burial, rearrangements to a more efficient packing are feasible; further, the intense pressure at those few angular grain contacts which are "carrying the load" causes local deformation or fracturing at these contacts until the load becomes more equitably shared. Since a seismic wave is transmitted from particle to particle primarily at the grain contacts, we are intuitively prepared to accept that considerable change of seismic characteristics—particularly velocity—may accompany these adjustments to the burial process.

The behaviour of the sands is further complicated by **cementation**.

In the case of the well-sorted well-rounded sands, it is fairly easy to see what must happen: the porosity must decrease, the proportion of solid matter to fluid matter must increase, the density must increase, the rock must become harder, and the velocity must increase. All these changes are likely to occur progressively; just like staying at a Holiday Inn, there are no surprises.

For an angular sand, however, the initial deposition of a hard cement at the angular contacts (and near-contacts) transforms the rock into something much more like a well-rounded sand.

We see two agencies, therefore, which can cause an unexpectedly large dependence of velocity on depth in an angular sandstone. One is a rearrangement, a tighter packing, of the angular grains; this occurs under the influence of depth alone. The other is the removal of the odd behaviour associated with differently-stressed angular contacts, by the growth of solid cement around and between the grain points; being dependent on the cementation, this effect often increases with depth also.

The constituents of the total picture are these:

- the basic behaviour of a well-rounded well-sorted sand, showing comparatively small dependence of velocity and density on depth,
- superimposed on this, a progressive and smooth increase of velocity and density with the degree of cementation,
- the anomalous behaviour of an angular poorly-sorted sand, showing abnormally low velocity at shallow depth,
- superimposed on this, the dramatic effect of even a little cementation, in bringing the angular sand much closer to the rounded sand.

The effect of depth on a sand is therefore not as clear-cut as the simple compacting effect of depth on a shale. There **is** a dependence on depth in a sand, but it is generally smaller than in a shale unless the sand is angular and poorly-sorted. The dependence on cementation is likewise variable, and this dependence is further complicated by the fact that cementation is a function of the volume of water migrating through the sand, the minerals dissolved in that water, the pressure gradient, the temperature gradient, the age, the presence or absence of hydrocarbons, and many other factors.

If we need some simple generalizations, then, they can be no more precise than these:

- very high values of density and velocity in a sand must signify low porosity,
- low values of density must signify high porosity, and
- low values of velocity may signify high porosity or unconsolidation (or both).

Figure 2.2.2–1 attempts to illustrate the situation for a high-porosity sandstone and a lower-porosity sandstone (porosities of 20% and 5%, respectively, at 4000 m (13000 ft)). The densities show a very simple (but less marked) variation with depth; this increase is in part the effect of increasing cementation and in part the (minor) effect of compaction. The velocity band suggests a

FIGURE 2.2.2–1

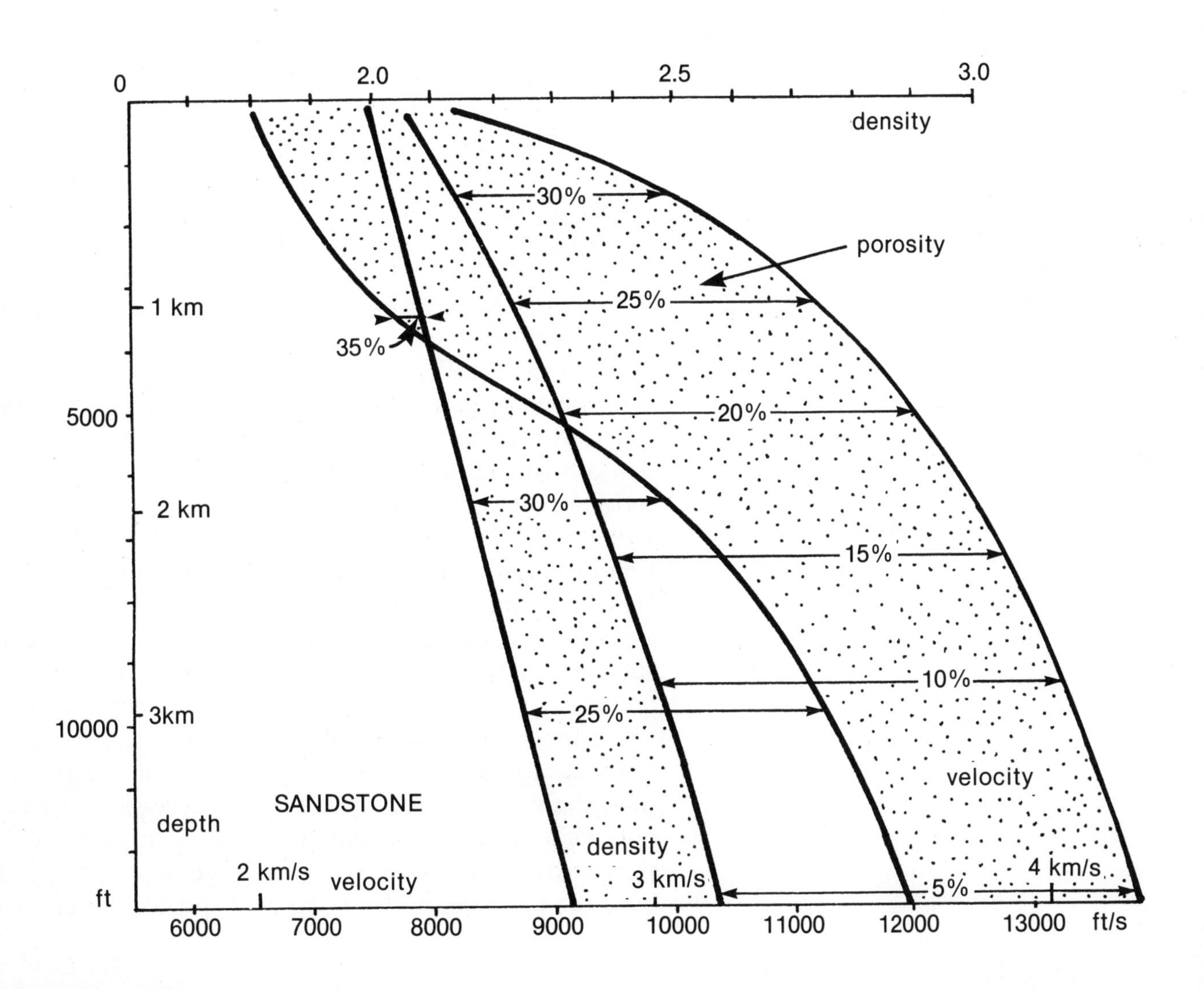

range of conditions: the high-velocity side represents a beach or dune sand with some cementation present even at shallow depth, while the low-velocity side represents an angular sand which is unconsolidated at shallow depth but in which cementation becomes significant at medium depth.

The porosity values are marked on the lines joining corresponding density and velocity curves. For the same porosity, the density is about the same for a sandstone and a shale, but the velocity is higher for the sand. As we might expect from the above discussion, the well-rounded well-sorted sand shows a progressive increase of velocity with depth, this variation actually being due mostly to the cementation as it reduces the porosity. The angular sand shows a strange dog-leg in the velocity curve; over the depth range where the packing improves and the cementation annuls the effect of the angular contacts, there is a dramatic proportional increase in the velocity.

Figure 2.2.2–2 uses the data of the last two figures to display typical values of acoustic impedance as a function of depth. The curve for the high-porosity sandstone retains the strange dog-leg associated with the consolidation process, and is generally lower than the shale curve. The curve for the lower-porosity sandstone is much higher than the shale curve.

Now we can see clearly the reason for some of the unwelcome features evident on seismic sections from sand-shale sequences. We recall that the strength of a shale-to-sand reflection depends on the contrast of acoustic impedance. First we see that the contrast between the shale and the **lower**-porosity sand is very marked, being sufficient to generate a good positive reflection. (Although the contrast appears to increase with depth, the reflection coefficient—and so the reflection amplitude—is actually constant in this example.) However, we also see that the contrast between the shale and the **high** porosity sand is much smaller, being sufficient to generate only a very weak reflection. The reflection is virtually zero at depth; it is likely to be visible only in the range of the dog-leg, where it is significantly negative. In this case the shallow part of a sand-shale section would show as a succession of weak but visible reflections, while the deeper part would show virtually no reflections; this is a common observation, for example, in the Tertiary of the northern North Sea.

Just a little more cementation in the high-porosity sand of this figure would eliminate the shale-to-sand and sand-to-shale reflections at all depths. It would still be an excellent reservoir. We are forced to concede that *many excellent sandstone reservoirs in a sand-shale sequence do not show on seismic sections*. It is the poorer ones which are likely to show.

Several factors may modify these conclusions to some extent:

- Locally, the region of a shale close to a good sand is likely to be more compacted than elsewhere (SITPA 60, 582). This moves important parts of the shale curve to the right; it may increase the visibility of high-porosity sands—but only at the expense of medium-porosity sands.
- The presence of **oil** in the reservoir, instead of the water assumed above, has the effect of moving the

FIGURE 2.2.2–2

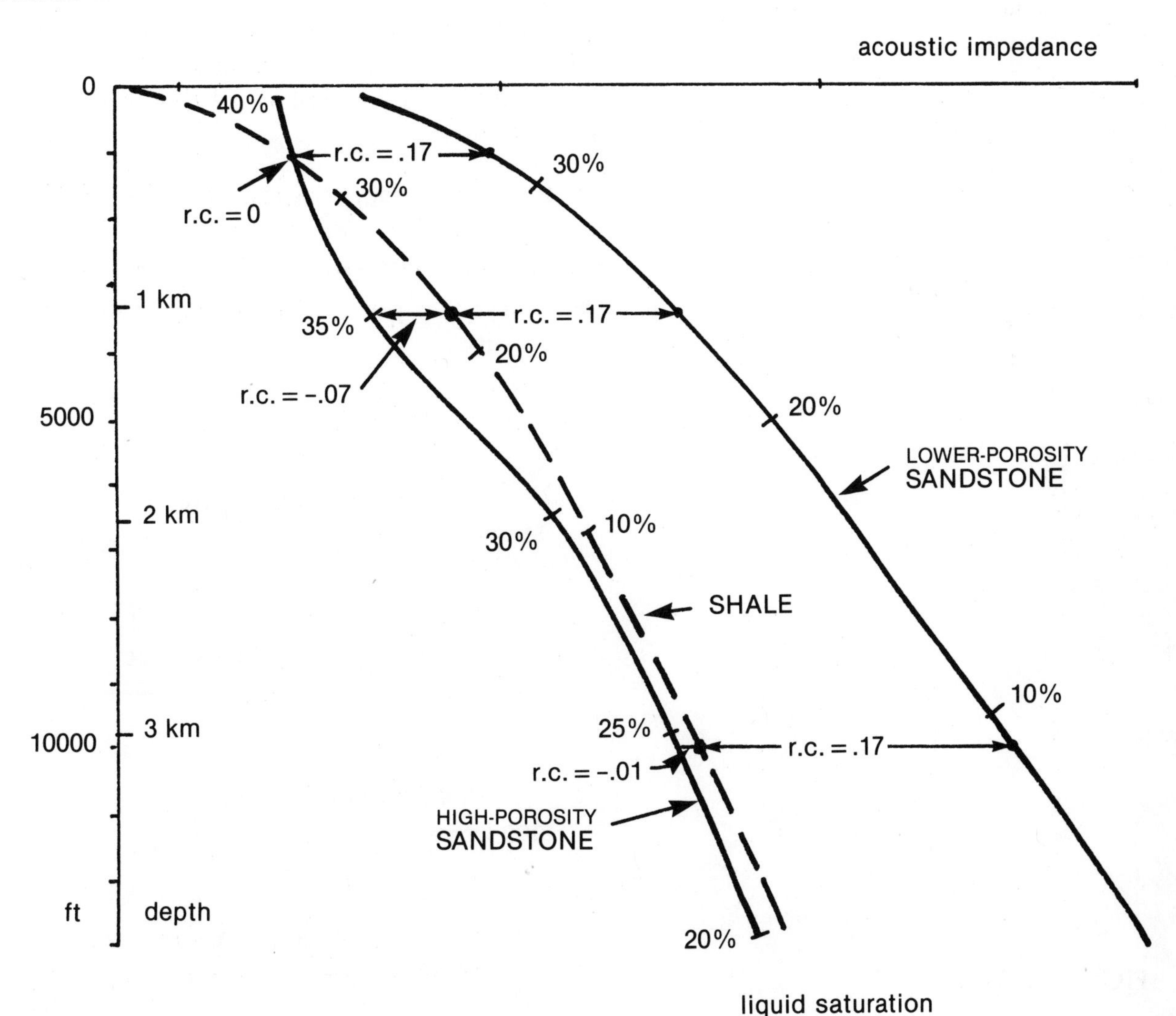

sand curves slightly to the left; this again may marginally increase the visibility of high-porosity sands.

- The presence of **gas** in the reservoir has the effect of moving the sand curves very considerably to the left (SITPA 94–100, 112). This may even bring the lower-porosity curve close to the shale curve, thus making the high-porosity sands very visible indeed (as bright spots) at the expense of the lower-porosity sands.

Whatever happens, however, we cannot escape from the unwelcome fact that the acoustic impedance of shales tends to be between the acoustic impedance of a very good sand reservoir and that of a poor sand reservoir; consequently there will always be a value of porosity—whatever the depth, and whatever the saturant—for which the sand is invisible.

The general situation is illustrated in figure 2.2.2–3. The details will vary, depending on a host of factors, but the broad conclusion remains.

However could we explain all this to the man in the street? Let us try.

- It is no great surprise that sandstone is harder than shale—that is one reason why it makes a better building material.
- Unless it is porous; the hardness decreases if it is porous.

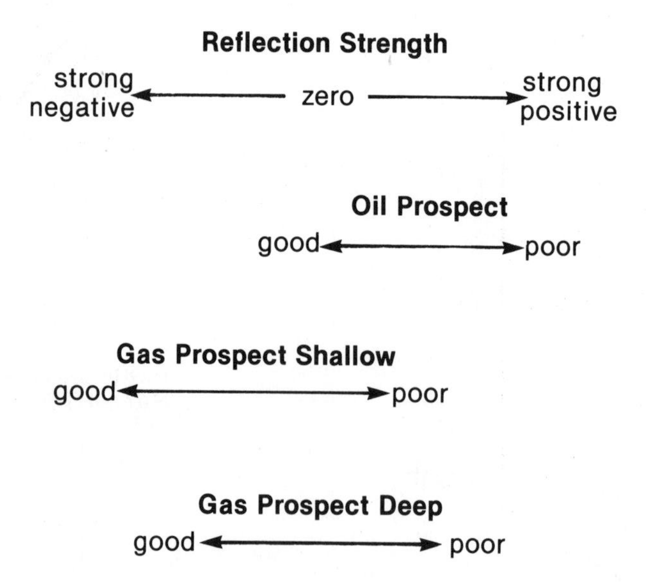

FIGURE 2.2.2–3

- In our sense of hardness (which is largely concerned with resistance to deformation) it is easy to see why the hardness depends on the porosity. If a rock has no porosity, it can deform only by compression of the rock grains themselves—which is tough. As soon as it has porosity, the grains can deform into the pore space; this is easier, because liquids are easier to compress than rock.
- If the pores contain gas—even just a little gas, with the rest liquid—there is virtually no resistance to the deformation of the rock into the pore space, and so the rock appears soft (in our sense).
- So it is sandstone of medium-to-poor porosity which is harder than shale. With good porosity it may become much the same, and with some gas it may even become softer.
- Now considering seismic reflections, we know that the strength of a reflection depends on the contrast of hardness. That is why we get a better echo from a stone wall than from a frame wall.
- There is a good (positive) contrast of hardness between shale and non-porous sandstone. So we get a good reflection from a bad situation.
- There is a good (negative) contrast of hardness between shale and very-porous gas-saturated sandstone—at least at shallow depths. That is why we get bright spots, and a bright spot is a good reflection from a good situation. But we see the caution: just a little gas, with a lot of water, gives **as good** a reflection—but from a poor situation.
- In between (where most of us live, most of the time), fair-to-good porosity filled with liquid or medium-to-fair porosity filled with gas may bring the hardness of the sandstone close to that of the shale, and so give no detectable reflection from a fair situation.

Thus a weak shale-sand reflection which locally becomes strong may mean a change from fair porosity to no porosity (which is bad), or it may mean a change from fair liquid-saturated porosity to fair or good gas-saturated porosity (which is good). We can resolve this only with reflection **polarity**.

Only for gas is there a fairly direct connection between the strength of the reflection and the appeal of the reservoir. Perhaps we should school ourselves to talk about **strong** reflections rather than **good** reflections, since in exploration for oil sands a good reflection is bad.

2.2.3 Carbonates

At any given porosity, shales and sandstones have about the same density, but carbonates are markedly more dense (SITPA 55).

At zero porosity, the velocity of a carbonate is higher than that of a sandstone, and consequently even higher than that of a shale.

Therefore we get a good reflection from the boundary between a sand and a totally-cemented carbonate, and an excellent one from the boundary between a shale and the same carbonate.

But again there are complications when we allow porosity.

The first complication concerns the manner of variation of primary porosity with depth. Usually, a significant part of this variation is contributed by compaction (as in shales) and a significant part—probably the dominant part—by cementation (as in mature sandstones). The overall effect is of a velocity-depth variation similar in degree to that of shales. This, then, is very acceptable; at least we are assured of a fair or good reflection from both a shale-lime and sand-lime interface, whatever the depth.

The second complication concerns the secondary porosity often observed in carbonates. Fractures can produce a significant depression of velocity—without much porosity or much change in density. A profusion of vugs can produce a significant depression of velocity in the bulk material, which may not appear on sonic logs. And so on.

However, because we are not concerned in this course with carbonates as reservoirs (and because we need every simplification we can get, after the last section), we shall assume from now on that all our carbonates are hard and tight. We shall consciously turn a blind eye to all situations where a carbonate made soft by porosity is in contact with hard sand—situations where our expected fair-or-good positive reflection would become weak, or zero, or even (in the case of gas) negative.

The appeal of carbonates, in exploration for sandstone reservoirs, lies in their value in generating marker reflections—often strong, continuous (until there is good reason for loss of continuity) and discrete. These are the reflections we like to use for correlations; these are the reflections we use for datumizing, and for mapping intervals; these are the reflections into which we hope our sand-filled channels and submarine canyons are cut; these are the reflections we use for calibration of reflection coefficients.

2.3 Implications for the strength and polarity of reflections

All of the foregoing material is summarized in tables 2.3–1 and –2. Of course these tables are grossly simplified; they are reduced to three rock types, and they ignore a thousand modifying factors. However, in anything as complicated as the earth we have to simplify to see any order at all.

TABLE 2.3–1: Reflection Strength and Polarity—SHALE TO SANDSTONE

Depth	Porosity of sand	Saturant in sand	Reflection strength	Reflection polarity
shallow	poor	liquid	strong	positive
		gas	medium	positive
	medium	liquid	medium	positive
		gas	weak	positive
	good	liquid	weak	positive or negative
		gas	very strong*	negative
deep	poor	liquid	strong	positive
		gas	medium	positive
	medium	liquid	medium	positive
		gas	weak	positive or negative
	good	liquid	weak	positive or negative
		gas	weak-to-medium	negative

SANDSTONE-TO-SHALE reflections, for the same situations, have the **same strength** but **opposite polarity**.

* Bright-spot conditions.

TABLE 2.3-2: Reflection Stength and Polarity—SANDSTONE TO CARBONATE

Porosity of sand	Saturant in sand	Reflection strength	Reflection polarity
poor	liquid	weak-to-medium	positive
	gas	medium	positive
medium	liquid	medium-to-strong	positive
	gas	strong	positive
good	liquid	strong	positive
	gas	very strong*	positive

CARBONATE-TO-SANDSTONE reflections, for the same situations, have the **same strength** but **opposite polarity**. No distinction of depth is made in this table, since depth has less effect than in the SHALE-TO-SANDSTONE case. The table assumes negligible porosity in the carbonate. The likely effect of porosity in the carbonate is to weaken all reflections, but not generally to change the polarity. It is conceivable that the interface between a tight sand and a gas-filled porous carbonate could be negative.

* Also bright-spot conditions, but geologically uncommon.

Table 2.3-1 summarizes reflection strength and polarity for *Shale—Sandstone* interfaces.

Table 2.3-2 does the same for *Sandstone—Carbonate* interfaces.

Suppose that, in a sand-shale sequence (and in the absence of other criteria), we see our top-reservoir reflection change from strong to weak. Where would we drill?

If we have well control, of course we would first make a comparison there. Otherwise:

- If we can establish that the reflection is negative we would drill the strong zone, and hope for gas.
- If we can establish that the reflection is positive we would drill the weak zone, and hope for oil.
- If we cannot establish the polarity we would drill the strong zone in a shallow gas-prone area, and the weak zone otherwise.

However, there are many complicating factors which we must take into account in practice, and the first of these is our next topic.

2.4 The problem of thin beds

A low-porosity liquid-saturated sand encased in uniform shale produces a positive reflection from its top and a negative reflection (of the same amplitude) from its base. A high-porosity gas-saturated sand in the same

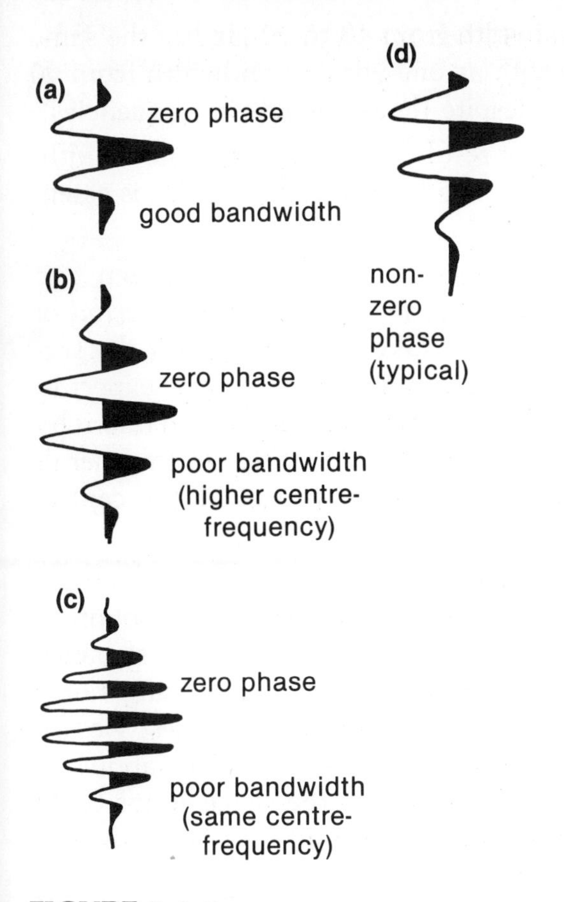

FIGURE 2.4–1

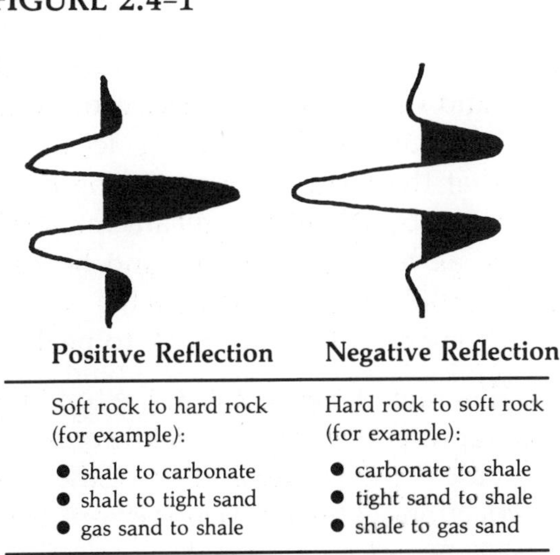

Positive Reflection	Negative Reflection
Soft rock to hard rock (for example):	Hard rock to soft rock (for example):
• shale to carbonate	• carbonate to shale
• shale to tight sand	• tight sand to shale
• gas sand to shale	• shale to gas sand

NOTE: The above convention for display polarity is adopted in this course, though it is at odds with the implications of the SEG recommendation on polarity (SITPA 538). The industry is increasingly accepting the polarity shown above, particularly because it alone is compatible with pseudo-log display.

All sections, whether commissioned from a processing house or received in trade, should be annotated to show the display polarity.

FIGURE 2.4–2

shale, at least at shallow depth, produces a negative reflection from its top and a positive reflection from its base.

If the sand is very thick, the top and bottom reflections are well separated, and we are able to see their opposite polarity. As the sand becomes vanishingly thin, the two reflections must ultimately cancel, giving nothing. So we explore what happens in between. In so doing, we can assume the basic strength-and-polarity expectations of table 2.3–1, and see how those expectations of reflection appearance are modified by the thinness of the reflecting layer.

First some generalities:

- The ability to distinguish the reflections from the top and bottom of a layer is called the vertical resolution.
- The vertical resolution depends on the *sharpness* of the seismic pulse; the sharper the pulse, the thinner the layer we can see.
- The sharpness of the pulse depends primarily on its bandwidth; the wider the band of frequencies it represents, the better (figure 2.4–1a and b). Good bandwidth is obtained—to limits imposed by filtering in the earth itself—by using a good source, by appropriate spread geometry and recording filters, and by deconvolution.
- The sharpness of the pulse also depends on its phase; the best condition is the zero-phase form (figure 2.4–1a and d). In this the pulse is symmetrical about a peak (black) for a positive reflection and about a trough (white) for a negative reflection (figure 2.4–2); to bring the pulse to this condition requires a phase-zeroizing step in the processing.
- We must understand clearly that the secret of resolution is **bandwidth**, and not just high frequencies. Of course, we can always raise the frequencies apparent on a section, merely by filtering out the low frequencies. But this does not improve the resolution; it just makes the pulse leggy, losing sharpness because of its profusion of cycles (figure 2.4–1a and c).
- For resolution purposes, it is useful to remember that the *effective* length of a pulse cannot be less than the reciprocal of its bandwidth. Thus a pulse

with a bandwidth from 10 to 20 Hz has the same effective length as one with a bandwidth from 90 to 100 Hz—despite the difference in frequencies.

- So the secret of resolving thin layers is bandwidth; phase zeroizing is a help, but bandwidth is essential.
- To study the effect of interference between two reflection pulses, we need the apparent *period* of the pulses. We can take this straight off the section, if we have one reflection which we expect to come from a discrete reflector, or we can find it by various other means (SITPA 213-219). In general, it approximates to the reciprocal of the centre-frequency of the bandwidth.
- Thus for a Vibroseis sweep of 10-40 Hz, which is a relatively poor bandwidth, we may obtain a *period* of as much as 40 ms. For the best bandwidth we can ordinarily hope for, using fairly small charges of dynamite in deep holes, we are not likely to see a period shorter than about 20 ms. Therefore we shall use the range 40 ms to 20 ms as representing the range of *periods* encountered in normal work.

Now let us use the classical Widess diagram, add some numbers to it, and summarize the discussion of it in SITPA 169–172, 193, 556-561. In so doing, let us be very sure to notice that the term λ (for wavelength) is used in a distinctive sense by Widess; he means the total duration of the pulse as he has drawn it, and he has drawn it so that the symbol λ represents **twice** the *period* of the seismic pulse (for example, from peak to peak).

- In the Widess diagram of figure 2.4–3, the top line (denoted 1 λ) represents a layer thickness large enough to provide complete separation of the top and bottom reflections. The pulses are zero-phase, and the two picks • and ○ represent the top reflection and the bottom reflection (in this case, negative and positive respectively.) The time thickness of the layer in this illustration is twice the peak-to-peak period of the top reflection (or the trough-to-trough period of the bottom reflection). Thus this time thickness is 80 ms for the 10-40 Hz Vibroseis example above, or 40 ms for the best-case example—a ratio of 2 to 1.

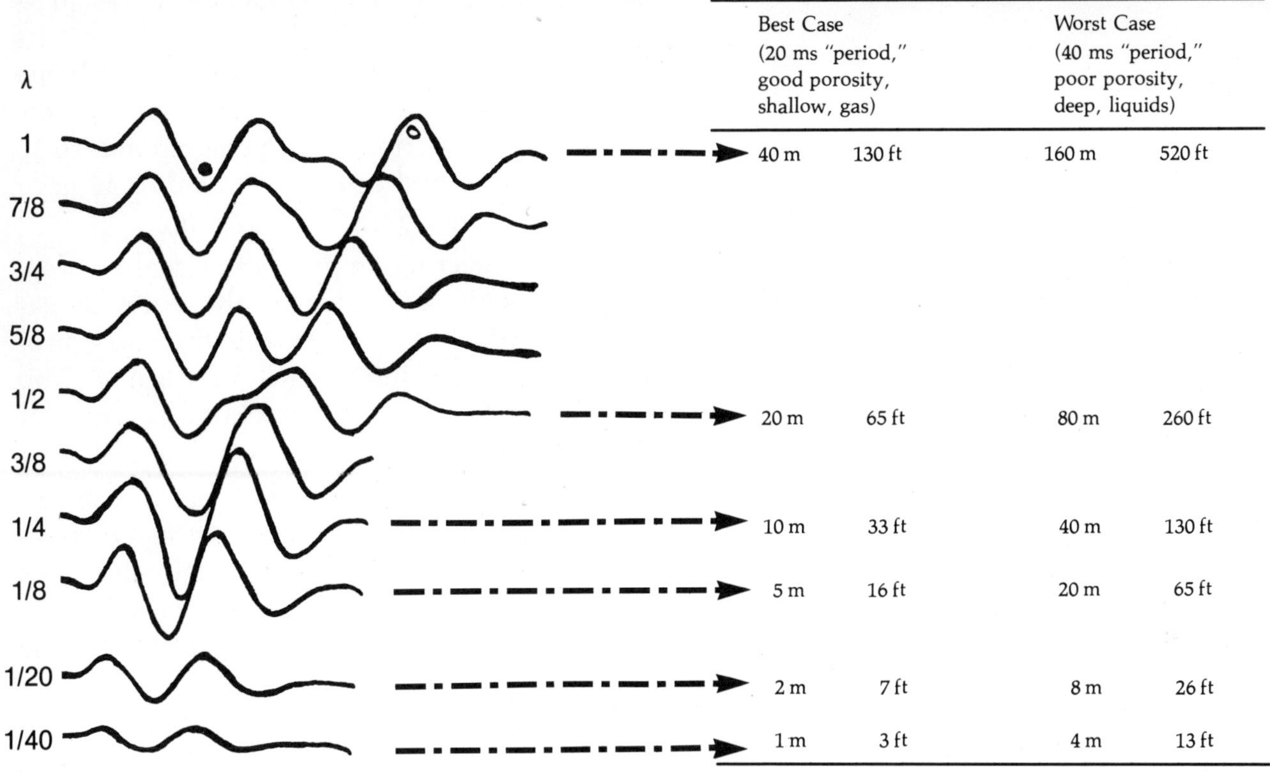

SAND RESERVOIR IN SHALE
(phase-zeroization applied)

λ	Best Case (20 ms "period," good porosity, shallow, gas)		Worst Case (40 ms "period," poor porosity, deep, liquids)	
1	40 m	130 ft	160 m	520 ft
1/2	20 m	65 ft	80 m	260 ft
1/4	10 m	33 ft	40 m	130 ft
1/8	5 m	16 ft	20 m	65 ft
1/20	2 m	7 ft	8 m	26 ft
1/40	1 m	3 ft	4 m	13 ft

NOTE: λ in the Widess illustration has the unusual meaning of the duration of the *complete* pulse (which is here equal to double the peak-to-peak period). Be aware of this point when making comparisons with other authors.

FIGURE 2.4-3 *(After Widess)*

- The 80 ms time corresponds to a thickness of 80 m at a low velocity of 2000 m/s, or 160 m at a high velocity of 4000 m/s (say 260 ft at 6500 ft/s, or 520 ft at 13000 ft/s); this is another ratio of 2 to 1. The best-bandwidth times represent half these thicknesses, so that we have a final ratio of 4 to 1 for worst-case to best-case. Thus a gas-saturated sand at shallow depths (where we might expect low velocities and best bandwidths) is totally resolved if it is 40 m or 130 ft thick. A liquid-saturated sand at great depth (high velocity, poor bandwidth) is totally resolved if it is 160 m or 520 ft thick.
- We should be so lucky.
- As the layer thins, the two reflection pulses interfere to form a complex. The complex is recognized as such—even though its components

are no longer clearly distinguishable—by its length, by saddles in its envelope, by hickies in its waveform, and finally by changes in its amplitude. There is a maximum of amplitude (denoted 1/4 λ) at a layer thickness of half the trough-to-trough period (which corresponds to 10 m or 33 ft in the best case above, and to 40 m or 130 ft in the worst case). This is *tuning*.

- Thus truly staggering amplitudes at shallow depths on a seismic section probably represent gas sands of the order of 10 m (33 ft) in thickness, providing constructive interference between the top and bottom reflections. The thickness is greater if the bandwidth is less than the postulated 20 ms.*
- At this thickness the pulse has become skew-symmetric (about a zero-crossing, instead of symmetric about a peak or trough as originally). This makes a useful corroboration, provided we couple it to the observation that *the period of the pulse has not changed*.
- For thinner layers, the reflection complex loses amplitude, and the only further change of pulse shape is a rise in apparent frequency (a shorter period).
- The key layer thickness in this context is 1/8 λ (corresponding to a thickness range of 5-20 m or 16-64 ft, depending on the bandwidth and velocities as above). At this thickness, the amplitude of the reflection complex remains larger than the amplitude of the top reflection alone, but the pulse shape has stabilized to a distinctive form.

  This form is characterized by:
  - a higher frequency than that of the top and bottom reflections,
  - one more half-cycle,
  - the skew-symmetric shape.
- We call this form **differentiated**. As noted above, the skew-symmetry indicates that the layer producing the complex is of thickness less than 5 m

* For example, the remarkable amplitude stand-out of the section on p. 24, in conjunction with the stated bandwidth of 10-40 Hz, suggests a sand thickness of about 20 m (67 ft) where the amplitudes are highest.

(16 ft) in our best case, or 20 m (64 ft) in our worst case; now the increase of frequency and the additional half-cycle (where visible) indicate that the layer is one-half these figures or less. The skew-symmetry is about an upward-going zero-crossing (polarity standard of p. 19) if the thin bed is soft (interesting) and about a **downward**-going zero-crossing if it is hard (less interesting, or even sometimes dangerous). Otherwise expressed, the largest amplitude is trough-to-peak for a porous gas sand, and peak-to-trough for a tight sand or a hard carbonate hoax.

- The last three lines of the diagram show no change of pulse shape. The only clue to the thickness of the layer is the **amplitude**.
- The penultimate line represents a trough-to-peak amplitude of about two-thirds that of the individual reflections from top and bottom. The layer thickness is 2 m or 6 ft in our best case, 8 m or 25 ft in our worst case. The last line still has about 40% of the individual reflection amplitude. The layer thickness is 1 m or 3 ft in our best case, 4 m or 13 ft in our worst case.
- In the case of bright-spot conditions (highly-porous gas-saturated sands at fairly shallow depths), the amplitude expected from the individual shale-sand interfaces may be so large that 40% of it still represents a strong reflection. Thus it is not unusual to see fair bright-spot indications from sands that are only a metre or two thick (say 3-7 ft).
- Deep in the section the sand-shale reflection amplitude is smaller, the velocities larger, the bandwidth is smaller—and so the thickness necessary for the layer to be visible is correspondingly greater. It may well be 10 times greater.
- Where the sand is thick, and the reflections from its top and base are measurable in isolation, we can make quantitative calculations of reservoir properties (SITPA 337-380). If the reservoir contains gas, and in favourable cases, we can calculate the thickness and the porosity, and get at least some idea of the water saturation. We are on the way to calculating reserves from seismic data. We shall return to this in later sections.

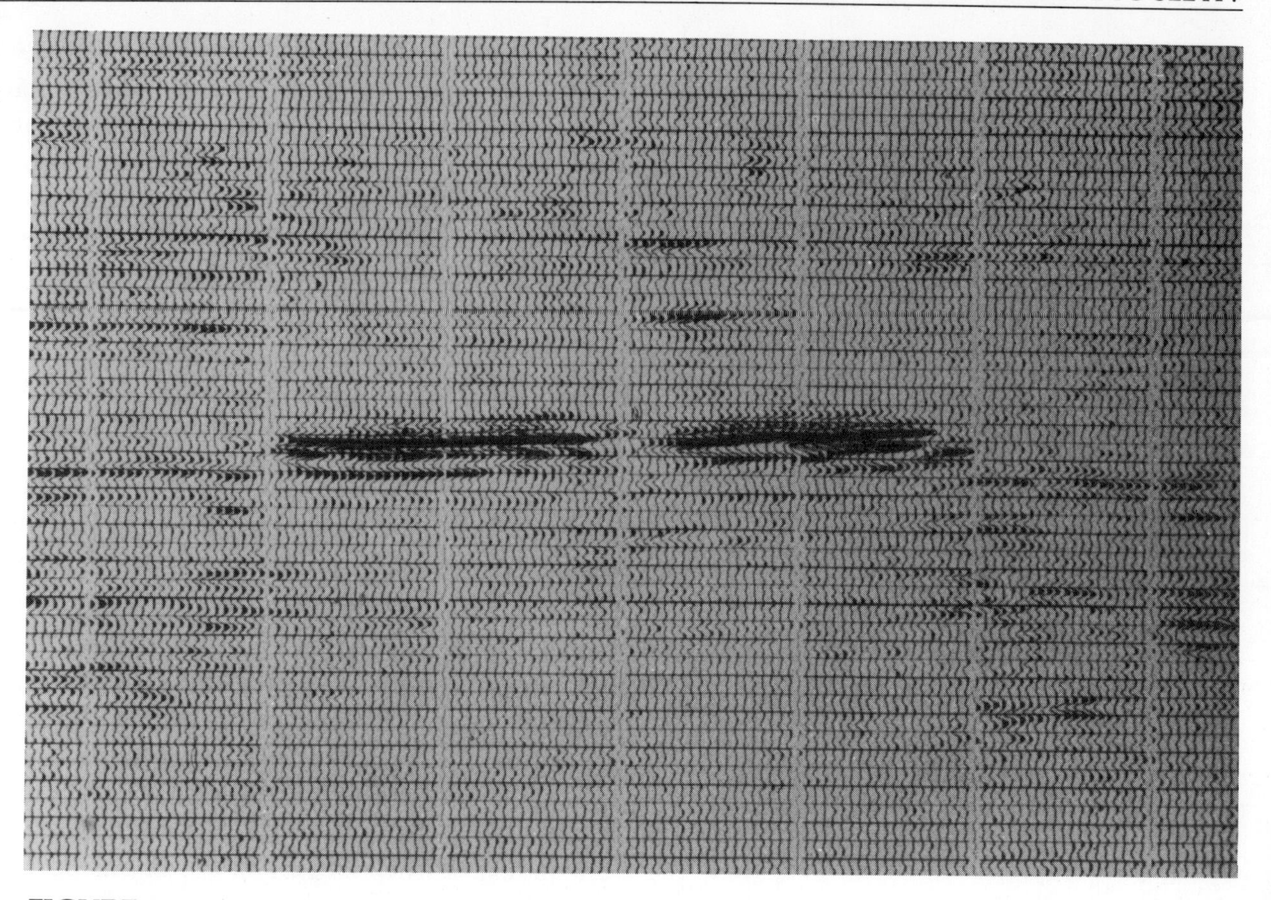

FIGURE 2.4–4 "Truly staggering" amplitudes in a Vibroseis bright spot (10-40Hz). *(Courtesy Prescott (Conoco).)*

- As the sand becomes thinner, in a pinch-out situation, a strong reflection alignment of spurious intermediate dip appears. This is particularly clear on the traces labelled 3/8, 1/4 and 1/8; it is one of the few exceptions to the principle we discussed earlier—that reflection alignments generally represent a time-stratigraphic horizon. This alignment is a hoax.

Beyond this point, as noted above, the shape of the reflection complex does not change, and no information can be gleaned on the thickness of the reservoir except from the fact that thinner layers yield small amplitudes. For the amplitude to be interpretable in terms of thickness, however, we must have a reference amplitude to which it may be compared. If we indeed have a wedge like the one illustrated, thick enough at one point to give separated reflections from top and base, then all is well. If not, we must calibrate the amplitude of some other reflection, and use that as a reference (SITPA 144-168, Lindsey et al., 1978).

The recognition of tuning on a seismic section is important to the interpreter, because it yields the time thickness of the bed. On one side of tune we expect to see the amplitude decrease and the frequency rise; on the other we expect to begin to see thickness suggesting two reflection components (figure 3.1-4).

- The Widess diagram used for all the above conclusions assumes the same material above and below the thin layer (and therefore equal amplitudes of the two reflections). This may be close enough in many sand-shale situations. For other situations we may have to draw an equivalent diagram, using reflection pulses of the same shape but different amplitudes. The general effect, of course, is a reduction in the appearances of interference—first constructive and then destructive—between the two reflections.

2.5 Transitions and transitional reflections (SITPA 172, 278–283)

Several geological mechanisms, both during and after sediment deposition, result in **gradual** changes of physical properties. In the present context, we are reminded of the fining-upward transition in point bars and the coarsening-upward transition in regressive sands. We now ask what is the effect on a reflection if the contrast of acoustic impedance is **transitional** (rather than abrupt, as heretofore assumed).

Perhaps, before proceeding, we should slip in a note about grain size and porosity. In some of this section, and in the next, we are implicitly accepting that coarser-grained materials suggest greater porosity. Of course this cannot be taken as a general truth; the total picture is much more complicated. In an ideal packing of uncemented spheres the porosity is independent of the grain size. In angular river sands the porosity usually increases with grain size, whereas in beach sands the reverse may be true. However, the effect of progressive cementation is usually to maintain porosity (and certainly to maintain permeability) in the coarser-grained materials. We therefore allow the implication here that coarser sediments are the more porous. Where this is not true the direction of an expected seismic variation may reverse; the details change, but we shall be able to see in what way we can expect them to change. In general, both porosity and permeability improve with better sorting, though again there are significant exceptions.

In figure 2.5–1, sketch 1 shows the basic reflector-reflection pair—an abrupt increase of acoustic impedance, and the positive reflection which it generates. For the reflection shape we assume a simple form, processed to the zero-phase condition.

Sketch 2 shows the corresponding negative pair: an abrupt decrease of acoustic impedance, and the negative reflection (of the same shape) which it generates.

The next two sketches are of thin layers (the condition of the Widess diagram) and their corresponding differentiated reflection shapes; sketch 3 would represent a tight liquid-saturated sand encased in shale, and sketch 4 a shallow and porous gas-saturated sand in shale (both for the polarity convention of p. 19).

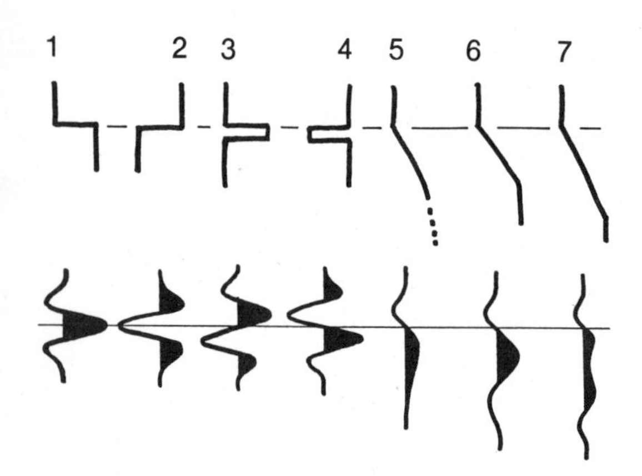

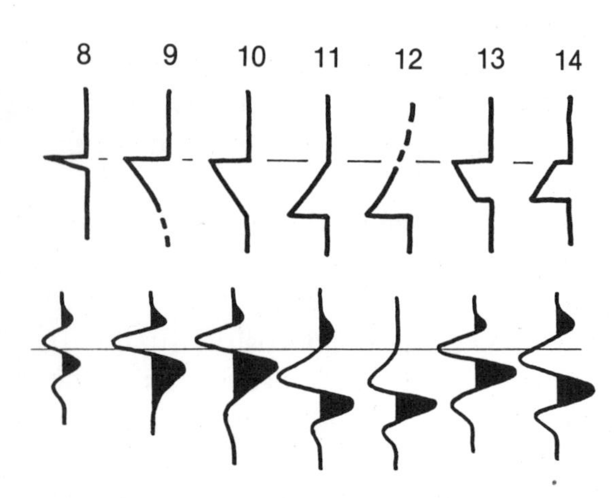

FIGURE 2.5–1

The other sketches represent various transitional situations. For simplicity, only one polarity is shown; the other follows by simple inversion, as in the first sketches.

Sketch 5 illustrates a shale overlying a massive liquid-saturated sand which is highly porous at its top and progressively less porous below. The reflection is weak, and of **integrated** form—**low frequency**, one fewer half-cycles, and skew-symmetric in the opposite sense to a thin-layer reflection.

Sketches 1, 3 and 5 (the step, the thin bed, and the slow transition) are of particular importance; the geophysicist can use these as elementary building-blocks, to synthesize at his desk the appropriate seismic response to any given geology.

Sketch 6 illustrates a thick liquid-saturated sand, of good porosity at the top grading into poor porosity at the base, and sandwiched between materials which happen to provide a match of acoustic impedance at both lithologic boundaries. The reflection is composed of two parts like sketch 5, with the second inverted. The overall effect (seen by comparing sketch 1 with sketch 6) is that replacement of an abrupt contrast of acoustic impedance by a transition of acoustic impedance, between the same limits, yields a low-frequency or **stretched** version of the reflection pulse.

If the transition extends over greater vertical distance, the two components separate, as in sketch 7.

In sketch 8 we consider the equivalent of sketch 4 (a thin porous sand saturated with gas) when the sand includes a coarsening-upward transition. There is no change of reflection shape, but a marked diminution of amplitude; therefore we cannot distinguish seismically between a thin sand with a transition and a thinner sand without a transition. This is not as bad as it sounds, because it means that reflection amplitude, in this case, remains a fair measure of the appeal of the sand as a reservoir.

Sketch 9 illustrates an upwards transition from shale to porous gas-saturated sand, capped by shale; thus it might represent a regressing shoreline subsequently inundated by a major transgression. The shale-to-gas-sand reflection is still seen, much as expected from sketch 2, but it is reinforced at the end by a low-frequency tail.

Sketch 10 is a counterpart to sketch 8, the sand becoming much thicker. We would detect the presence of the transition now, by the low-frequency tail; it would not be possible, in practice, to distinguish between this situation and that of sketch 9, but that would not matter much.

In sketch 11 the thick sand is coarsening downwards instead of upwards. We are probably ready to accept immediately that the reflection shape is reversed in time—with the low-frequency part first—but we might have to think for a moment to rationalize the fact that the reflection is also inverted in polarity.

So if we can visualize a thick sand which is coarsening upward in one place and fining-upward in another, we would see a sort of **Z** on the reflection section between them. The more-or-less horizontal bars of the **Z** might be a particular trough above the sand and a particular peak below; the appearance of dip between them would be the effect of the pulse changing internally from high to low frequency. If we insisted on forcing a pick on a high-amplitude trough, without seeing the significance of the changing pulse shape, we would infer dip—quite wrongly. However, the dip **does** show the attitude of the coarse component of the sand.

Sketch 12 is the upwards-downwards mate of sketch 9. Again we notice the low-frequency part—the signature of a transition—coming before the normal-frequency part. And again we note that corresponding parts have opposite polarities in sketches 12 and 9.

Sketches 13 and 14 represent coarsening-upward and fining-upward conditions in a sand of medium thickness. We might just be able to see the high-to-low-frequency change in sketch 13, and the low-to-high in sketch 14. Again a lateral change from coarsening-upward to fining-upward would yield a spurious impression of dip.

Although sketches 8-14 are shown for the case of a shallow and porous gas-sand, each sketch obviously has a counterpart—of opposite polarity—for the case of a tight liquid-saturated sand.

From all of this, we should select for emphasis a few welcome simplifications, and an unwelcome one:

- Transitions of acoustic impedance generate low-frequency reflections.
- A porous gas-saturated sand with a vertical gradation to poorer porosity contains less gas than if the whole sand had its maximum porosity; it also returns less reflection amplitude. The bright spot looks less bright—which is correct.
- A porous liquid-saturated sand with a vertical gradation to poorer porosity contains less liquid than if the whole sand had its maximum porosity; it also returns greater amplitude—which again is correct.
- In a sand-shale sequence, our greatest interest attaches to reflections which are strong and **negative**—in hopes that these relate directly to gas-saturated sands of good porosity. Our second interest attaches to reflections which are **weak** (and either positive or negative)—in hopes that these signify gas-saturated sands of poor porosity or liquid-saturated sands of good porosity. We have much less interest in reflections which are **strong and positive**—these are likely to indicate liquid-saturated sands of poor porosity. So the distinction of polarity is important; unfortunately, however, all changes in sketches 3-14 complicate polarity identifications—even with phase-zeroized pulses. It is true that sketches 8-14 (all gas sands, with various transitions) share with sketch 4 (the prototype gas sand) the fact that the maximum extremum-to-extremum amplitude is trough-to-peak, and this is a gratifying observation; however, sketch 5 shows the same property, so to

make a valid judgement we would need to add a second property (such as the number of cycles in the pulse).

Phooey.

2.6 Lateral changes of properties

So much for vertical changes of properties; now we must ask about the horizontal ones. How quickly can a sand change into or be replaced by another lithology, and what is the effect on the seismic section?

First let us establish that we do sometimes see, on seismic sections, reflections which maintain a constant strength (and so indicate a constant contrast of hardness) over long distances. These are often our marker reflections, on which we rely for structural picking. However, the best of these tend not to be sandstones, because of the lateral variation to which many sandstones are inherently liable. Even if a thin sand is not completely local, its thickness usually varies (so that the interference pattern between the top and bottom reflections also varies) or its porosity and its acoustic impedance vary (so that the strength of the component reflections also varies).

Next let us accept that if the contrast of acoustic impedance between two materials changes **slowly**, the reflection strength changes in proportion, faithfully.

However, a problem arises if the contrast of acoustic impedance changes **abruptly**. To understand why, let us put aside the naive idea of a "depth **point**" or "reflection point," and appreciate that the reflection process occurs over a **zone** of the reflector. (This is typical of geophysicists; they usually start with something comprehensible, and then immediately say, "But of course—it's not really like that.") An infinitesimal point could return only an infinitesimal reflection; what we need is a good extent of the reflector to participate in the reflection process. We shall call this extent the **reflection zone**; ordinarily it is a circle, about the nominal reflection point as centre.

The diameter of this circle depends on the depth of the reflector, the geometry of the spread, the velocity distribution, and the frequency content of the reflection. This is not as complicted as it sounds. We can calculate it with a simple equation, or—perhaps with a better physical feeling for it—we can just read it off an individual record.

Figure 2.6-1 is reproduced from SITPA 186, with the addition of the equation and some typical values. The diagram illustrates that the physical meaning of the equation is the distance over which the normal moveout increases by half a period. The typical values illustrate that the reflection zone is surprisingly large*—encompassing several or many so-called depth-points.

Now we can see the distinction between a gradual and an abrupt lateral change of properties. The reflection amplitude is a faithful** reproduction of the acoustic con-

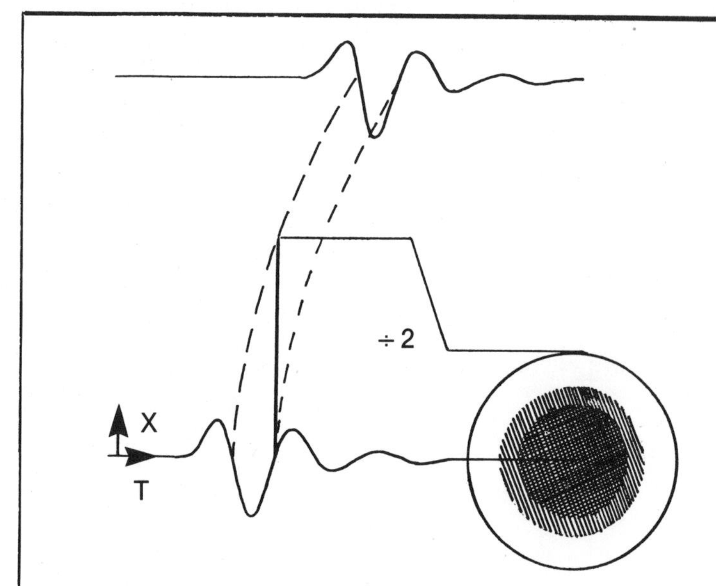

The diameter of the circle is approximately $V\sqrt{tp}$, where V is the rms velocity, t is the reflection time, and p is the peak-to-peak period of the seismic pulse. The *effective* diameter (the shaded zone) is about ½ of this.

Typical values for the effective diameter of the reflection zone:

shallow, in a low-velocity section ..... about 120 m (400 ft)
shallow, in a high-velocity section .... about 210 m (700 ft)
deep, in a low-velocity section .... about 750 m (2500 ft)
deep, in a high-velocity section .... about 1200 m (4000 ft)

FIGURE 2.6-1

* Some ornery geophysicists refuse to accept that the reflection zone is as large as this. They hold that some mechanism (as yet unknown) reduces the effective diameter to perhaps one-quarter of $V\sqrt{tp}$. However, everyone accepts that there is a reflection zone of appreciable diameter, and our discussion can proceed on this basis.

** For ideal recording and processing.

trast if that contrast does not change very much over the dimensions of the reflection zone; it represents some reasonable average if the properties change significantly but gradually over these dimensions; and it is quite false if the properties change abruptly within the zone.

When the properties change abruptly within the zone, **diffractions** are generated from the edges of the abrupt change.

Thus if the seismic line approaches a major fault (figure 2.6–2), the amplitude of the reflection, here suggested by the boldness of the event, starts to decrease as the reflection zone impinges on the fault. When the line reaches

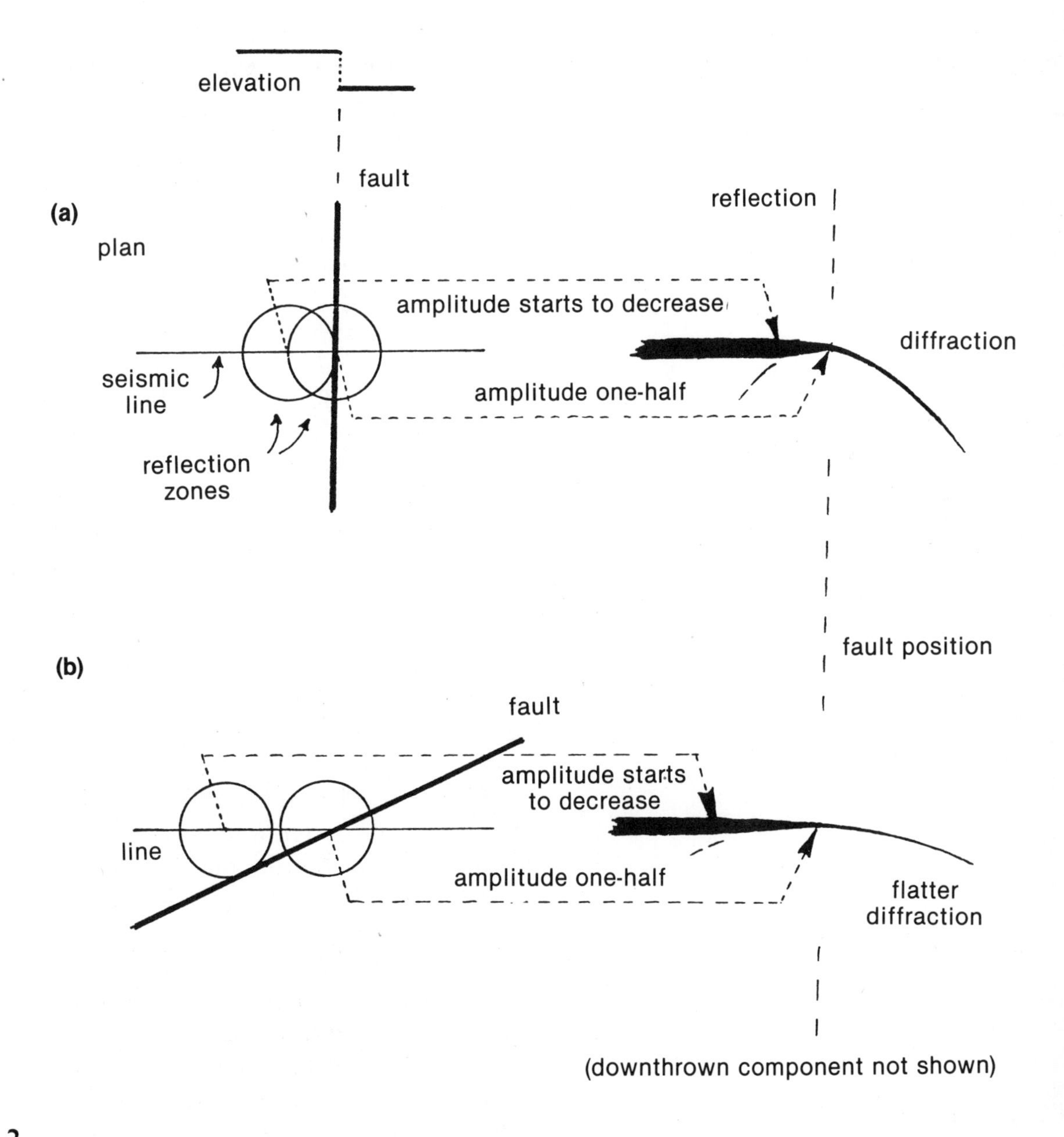

FIGURE 2.6–2

the fault the amplitude falls to one-half. Thereafter we see a diffraction. Since the curvature of the diffraction hyperbola is related to variables much the same as those which govern the dimensions of the reflection zone (SITPA 182-202, 285-288), the diffraction shows marked curvature when its origin is shallow in a low-velocity section, and much less curvature when its origin is deep in a high-velocity section. If the line is not perpendicular to the fault (lower diagram of figure 2.6–2), the loss of amplitude begins earlier, and the diffraction is flatter.

Figure 2.6–3a illustrates what happens as the line crosses a wide steep-sided channel sand, at right angles. Let us suppose the sand is shallow and gas-saturated, so that even if it is not very thick we still get a significant anomaly of apparent reflection strength. The reflection complex has its maximum strength only over the central part of the channel, where the reflection zone is contained within the width. On each side of this there is a gradual diminution of amplitude. The reflection anomaly is wider than the channel by the diameter of the reflection zone. Only the apices of the diffractions (if visible) give any indication of the actual position of the edges of the channel.

In figure 2.6–3b the channel width is smaller than the reflection zone. Therefore the amplitude never attains the value we would expect from the acoustic contrasts.

In figure 2.6–3c the narrow channel is shown inclined to the seismic line. The maximum amplitude is the same as in the last illustration, but the build-up starts from further away. Also, the diffractions are flatter.

We should note in particular that if the seismic line actually follows the axis of the channel, the amplitude is still low—still reduced by the proportion of the circular reflection zone which lies outside the channel. In this case, of course, we observe no diffractions.

Figure 2.6–3d modifies the abrupt edges of the former channels to represent a valley-fill more realistically. By inverting it, we could make it look like a lenticular shoreline sand. In either case, the reflection complex must show the wedging interference patterns of Figure 2.4–3, and zero amplitude at the feather-edge of the sand. Further, if the thinning of the sand is gradual, we shall not see any diffractions.

(If the other distractions on a long flight are inadequate, one many even find oneself looking down at a particular geologic feature—this stream valley, or that

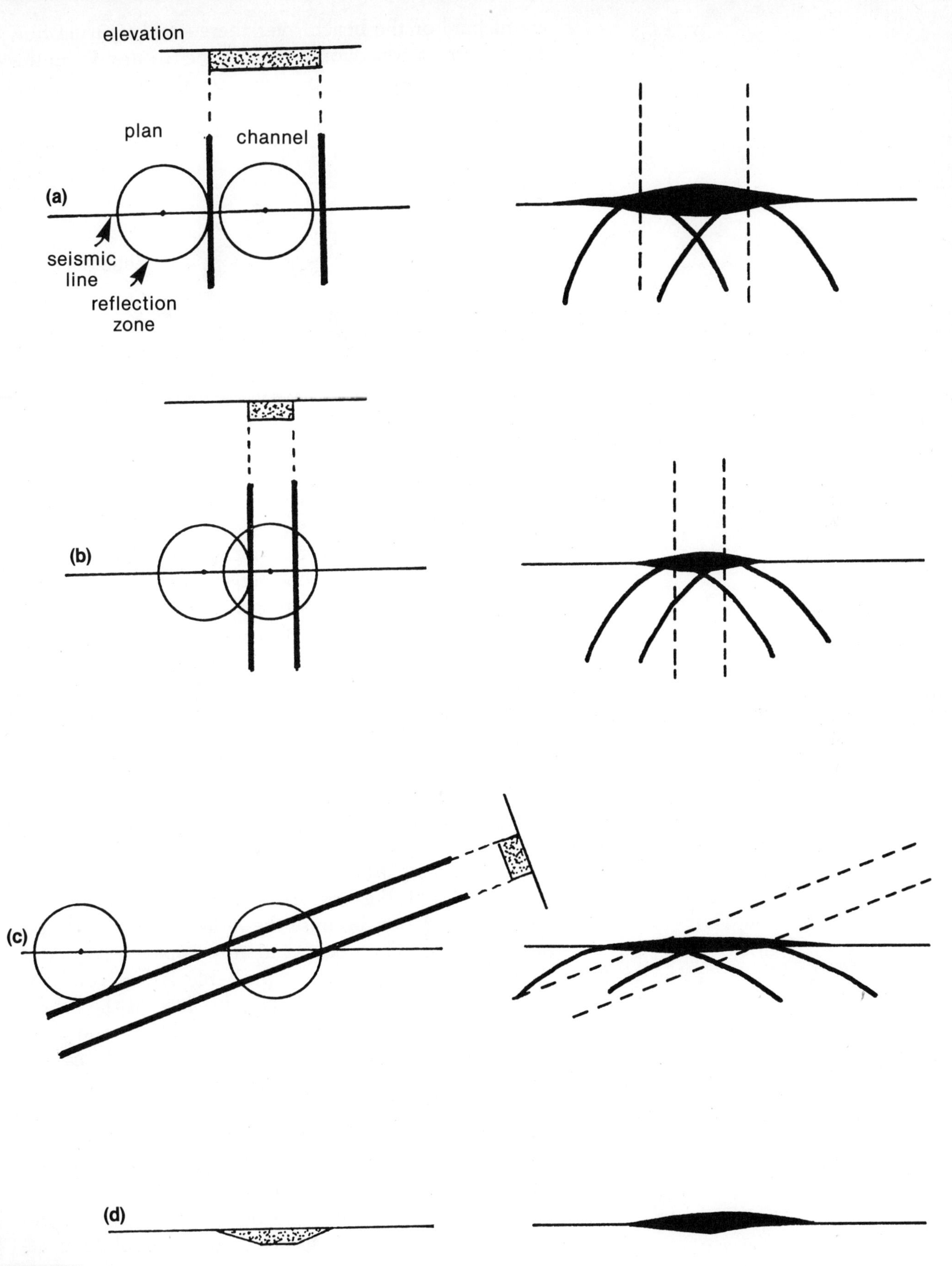

FIGURE 2.6–3

point bar, or the beach over there—and figuring how the reflection zone would fit over the feature. From this height, would it be detectable seismically?)

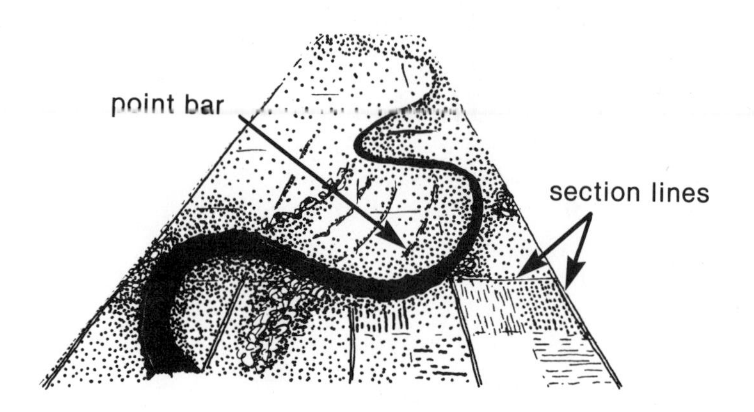

Let us consider the consequences of all this to the significance of amplitudes.

We have said that a shallow gas-saturated sand in shale must generate a **strong** reflection complex (with a negative top). Since the strength owes something both to the porosity and the thickness, we can say: the stronger the better. However, the strength also depends on the relation between the horizontal extent of the sand and the diameter of the reflection zone. The first consequence of this is that enormous reflection amplitudes must mean *not only that the sand is porous and thick but also that it is wide*. So we still say: the stronger the better. A second consequence is that a narrow sand must lose some of its amplitude, and so appear to have less porosity or less thickness, or both. A third consequence is that the degree of this loss of amplitude depends on depth, because the reflection zone is larger at depth; a good but local sand must appear less good at depth—not only because it probably is less good (by reason of cementation) but also because of the totally unrelated factor of the size of the reflection zone.

But at least we do have one simple rule, in the case of reflections from shallow gas sands: the stronger the better.

When we turn to liquid-saturated sands, however, this rule does not work. For in this case a **strong** reflection complex (positive top) means a **tight** sand. What we are hoping for is enough reflection amplitude to see, but no more; this, other things being equal, should indicate good porosity. But now the thickness and the lateral extent of the sand work in the wrong direction; a thin sand of small extent often looks better than a thick sand of infinite extent. In the study of liquid-saturated sands, this limits the usefulness of amplitude to cases where we can get some idea of the thickness and the extent by other means.

One other thing we should note—the effect of lateral variations in the cementation. Thus, from our earlier discussion of figure 2.6–2, we would expect significant diffractions from a major fault; however, this need not be so, and one of the possible reasons is variations of cementation. Near the fault, the sandstone components in the section tend to be more cemented than they are elsewhere; this effect, which is caused by the ready migration of mineral-rich waters up the fault plane, creates a lateral gradation of porosity which may extend over a significant distance. The sandstone therefore hardens in the vicinity of the fault. If this **increases** the contrast with an overlying or underlying shale (which is the case of liquid-saturated sands of medium porosity), we may expect to see marked diffractions from the fault at the sandstone level. If it **decreases** the contrast (which is the case of gas-saturated sands of good porosity) we may see no diffractions.

In practice, we have often been puzzled by very clear faults which show no diffractions; cementation is likely to be one of the answers, and fracturing near the fault (which works in the opposite direction, increasing the porosity) is another. The final outcome depends on the cementation, the fracturing and the recementation of the fractures—which should be sufficient to account for anything and everything we see on our sections.

Perhaps we should also have noted that the strength of our reflections depends not only on acoustic contrast, and thickness, and lateral extent—but also on structural curvature. However, for present purposes we shall ignore this additional complication; those for whom structural curvature might be a problem can find a discussion in SITPA 185-190.

2.7 Unconformities

Unconformities are the key to many important situations. Springing first to the mind are unconformity traps; however, the role of unconformities in defining sequence boundaries is perhaps even more important, since it affects **all** aspects of the search for sandstone reservoirs.

We have said that a seismic reflection comes from a contrast of acoustic hardness, and that the reflection mechanism is totally unconcerned with the geological circumstances creating such contrasts. There is no guarantee that a lithologic boundary or a time-stratigraphic boundary will produce a reflection whose amplitude is anything other than zero. Similarly there is no guarantee of a continuous reflection from a continuous unconformity.

Let us see what factors control the ability of the seismic method to detect an angular unconformity.

strong
stronger
soft
hard
harder

(1) First and foremost, as always, is the contrast of acoustic impedance. If we assume a uniform material above the unconformity, this means that the reflection must change in amplitude along the unconformity—as different truncated materials are brought into contact with the upper material at the unconformity surface. So the first characteristic of an unconformity reflection is **variable amplitude**—strong, then weak, then perhaps strong, and then perhaps zero. As if this did not make the picking sufficiently difficult, we must also note that—unless the material above the unconformity is either harder or softer than all the materials below—the reflection must also switch polarity back and forth along the unconformity surface. If we insist on picking a trough all the way along, we may be imputing to the surface relief which does not exist.

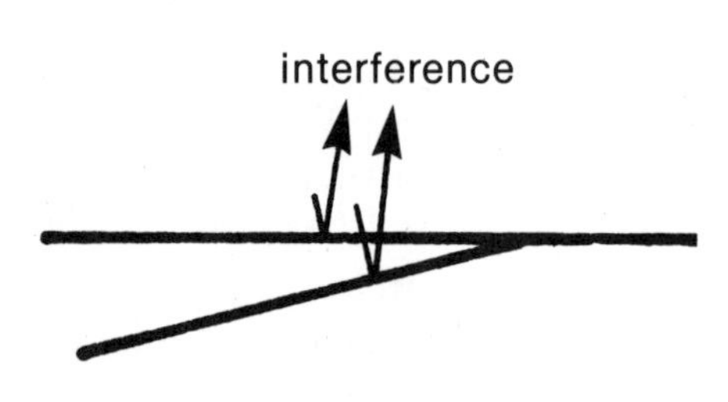

(2) Wherever the boundary between two subcropping formations is truncated at the unconformity, we have the conditions of the Widess diagram (figure 2.4-3). The reflections from the boundary and from the unconformity must interfere; for any practical seismic pulse, this means alternately constructive and destructive interference, so that the unconformity reflection is alternately strengthened and weakened. This effect is quite distinct in origin—though not always in appearance—from the hardness effect in 1 above.

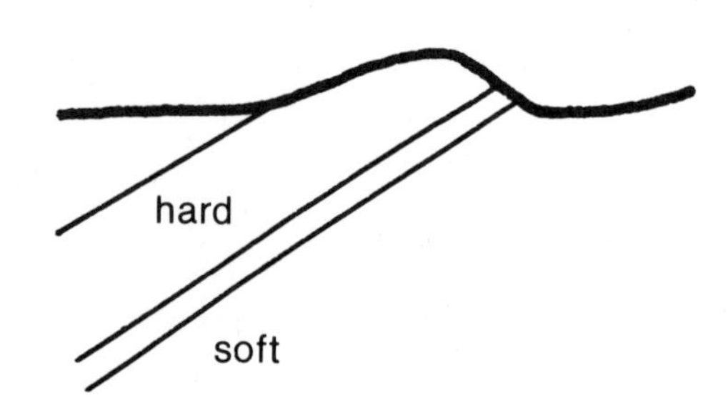

(3) Often, the unconformity surface is rough; the hard materials stand high, relative to those more easily eroded. The scale of this roughness is usually such as to decrease the amplitude to be expected from the hard (convex) materials (SITPA 185-190). The situation in the

lower reaches of the surface is less easily specified, particularly as they are likely to contain some of the products of erosion from the harder formations. What is clear, however, is that unless the truncated surface is substantially plane there must be additional fluctuations of reflection amplitude caused by roughness.

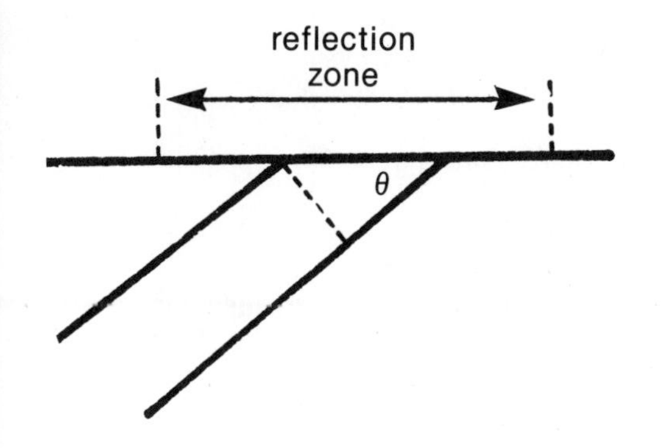

(4) Now we must superimpose on items 1 and 3 the effect of the circular reflection zone. For example, let us visualize what happens at the subcrop of a hard bed 50 m thick. Because it is harder than the material overlying the unconformity, we expect a strong positive reflection at the unconformity surface. Because the hard formation was probably a high ridge before inundation, we expect a weakening of this reflection on account of convex curvature. Because the width of the bed on the unconformity surface is likely to be less than the diameter of the reflection zone, we expect a further weakening; since the width of the bed is its thickness divided by the sine of the angle of the unconformity, this weakening is less for small unconformity angles. The effect of a large-diameter reflection zone is always to **blur** the amplitude changes we expect from the other factors affecting the amplitude.

(5) We have said that we expect diffractions wherever there is a rapid spatial change of acoustic hardness. At an angular unconformity, this condition must arise whenever two adjacent truncated beds are markedly dissimilar. In other words, if the interface between two formations is generating a strong reflection, and then those two formations are truncated at an unconformity, we expect significant diffractions from the point where the three materials meet. A sequence of strong reflectors, when truncated at a steep angle, generates the appearance of a thousand Fu-Manchu moustaches. This is so even without the extra complications of roughness—even if the unconformity surface is perfectly plane. With the roughness of erosional topography added, the diffractions may become the most obvious indication of the existence of the unconformity.

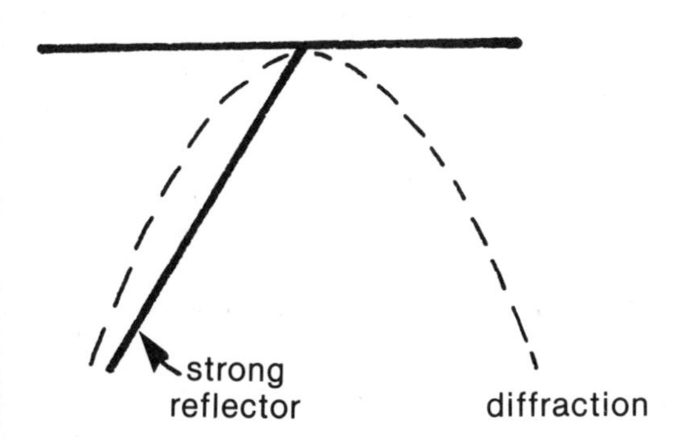

(6) Unconformities also generate unwelcome refraction effects. These are discussed in SITPA 190, 267, 457, and we shall not consider them further here.

From all this, it is clear that the picking of unconformity surfaces is fraught with problems—reflections continually changing shape, reflections switching polarity, reflections below the noise level over considerable distances, and sometimes a profusion of diffractions.

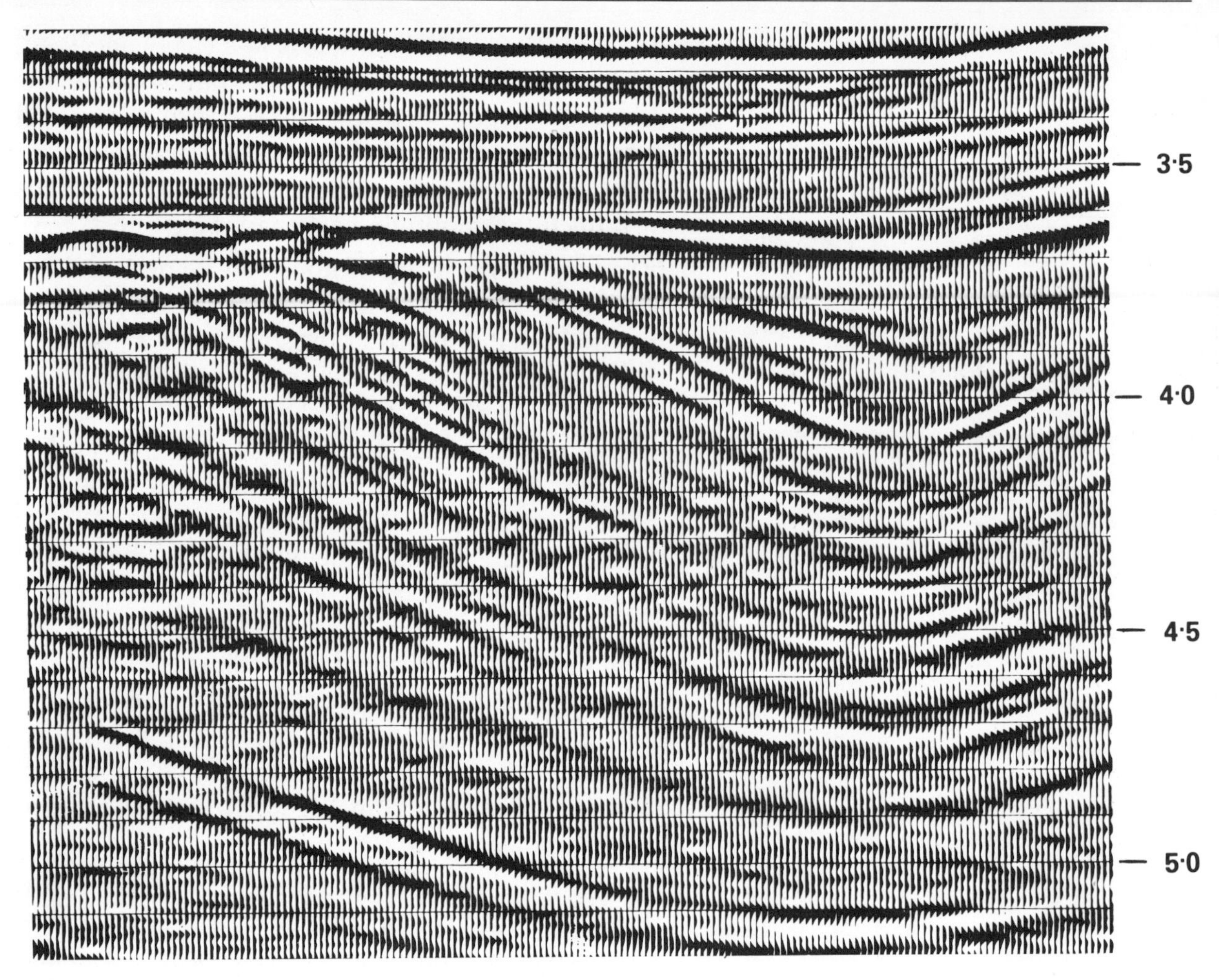

FIGURE 2.7-1 *(Courtesy S&A Geophysical)* **Exercise:** *Well, what do we think?*

Good continuous reflections are for kids; unconformities are for **men**.

Persons.

2.8 The value of edges

Since many sandstone bodies are local in their extent, we should explore what use can be made of their edges.

First, we shall see later that the **shape** of the body (in plan) is a useful guide to its type.

Second, there are some useful conclusions about the acoustic hardness that can be drawn from studying the reflection behaviour on and off the local body (provided that the body is large relative to the dimensions of the reflection zone).

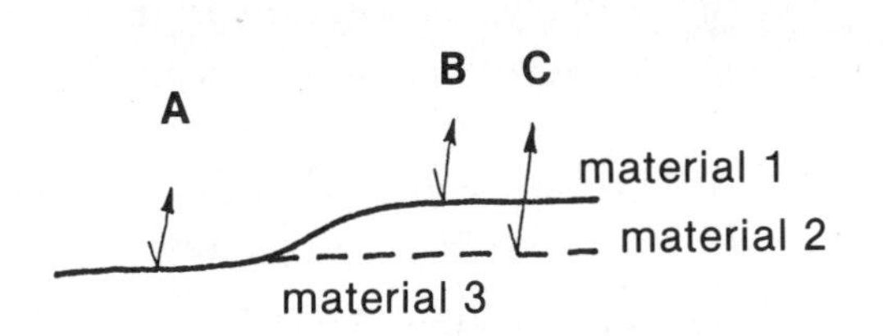

We sketch the acoustic-impedance log for two basic situations—

When reflector A is

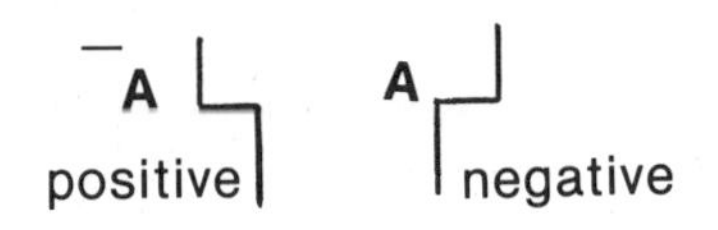

Then, on the sand body:

If the strongest reflection is A, the sequence is

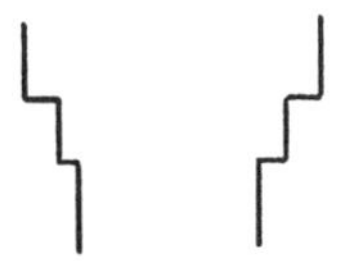

In this case all three reflections have the same polarity.

If the strongest reflection is B, the sequence is

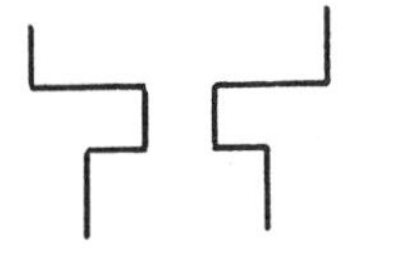

In this case A and B have the same polarity, and C is opposite.

If the strongest reflection is C, the sequence is

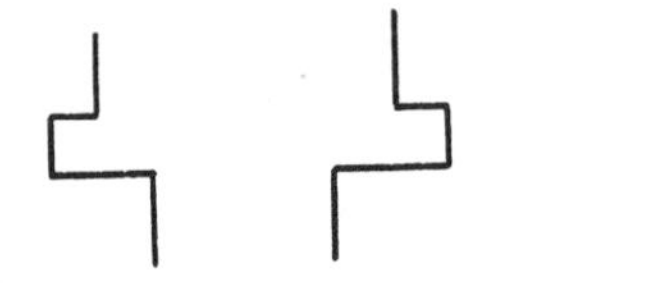

Now A and C have the same polarity, and B is opposite.

FIGURE 2.8–1

Let us postulate the situation shown in figure 2.8–1, in which material 2 terminates; then the overlying material 1 comes to be in direct contact with material 3. We have three reflections of strength A, B and C, as shown. Then what can we say?

If all three reflections are discrete, and the recording and processing are ideal, we can draw the conclusions given in the figure. For example, *if reflection A is the strongest, the hardness of material 2 must be intermediate between those of materials 1 and 3.* Further, the polarities of all three reflections must be the same, though we cannot say (from the strengths alone) what that polarity is. It could be that material 2 is a sand, between a shale and a carbonate; it is **most unlikely** that material 2 is a hard non-porous carbonate.

That is a fairly useful conclusion to follow from a very simple observation—just that A is strongest of the three reflections.

In the situation illustrated, reflection A is from the interface between two fairly massive bodies. It should therefore be straightforward to obtain its polarity—both from the pulse shape and (provided there is no salt or anhydrite involved) from the interval velocities. So if we have a good reflection at A, and it is positive, then the probable sequence is shale-sand-carbonate; if A is negative it is carbonate-sand-shale.

The ratio of strengths between reflections B and C tells us whether material 2 is closer in hardness to material 1 or material 3.

The second and third possible situations of figure 2.8–1 imply that *the hardness of material 2 must be either greater than or less than that of both material 1 and material 3.*

So we probably have **either** a porous gas-saturated sand or a hard carbonate. If reflection A is positive, it is the carbonate if reflection B is the strongest; it is the gas sand if reflection C is the strongest. If reflection A is negative, the implications are reversed.

This type of simple but useful thinking can be extended to a variety of situations.

Of course, problems arise sometimes. Perhaps the amplitude information is inconsistent; perhaps, for example, reflection A appears **equal** to reflection C—which is impossible on the assumptions we have made. Then the assumptions are wrong. If we have kept good control of the recording and processing, the problem must mean that there is a **transition** of hardness within material 2 (which might be very reasonable if it is a sand) or that at least one of the reflections is a thin-bed complex.

Or perhaps the amplitude relationships do not appear consistent with the observed polarities. Then again we know that the assumptions are wrong— which may still be useful information, obtained at very small effort.

3 The Seismic Signatures of Typical Sandstone Reservoirs

Now let us apply our previous discussions to the individual types of sandstone reservoirs.

We have already identified the factors which define the reflection response of a sand body. Let us divide them into three groups:

Group 1

This group identifies the factors decided by the **sediments** and the **depositional environment**.

- The **other** material. We put this first, because it is the one most likely to be overlooked; a reflection comes from the interface between **two** materials, and the overlying and the underlying materials are just as important as the sand.
- The original porosity of the sand (as a major component defining the acoustic impedance).
- The thickness of the sand (as a major component defining the interference pattern between the top and bottom reflections).
- Vertical transitions of porosity within the sand (as a major component defining the frequency content of the reflection complex).

- The area of the sand (in relation to the area of the reflection zone).
- The abruptness of the lateral limits of the sand (defining the generation of diffractions).
- The roughness of the surfaces of the sand.
- Possibly, cross-stratification in the sand (but only if the sand is thick enough, and the variations of hardness within it are significant).

Group 2

This group identifies the factors decided by **geological processes after deposition**.

- The saturant; here we accept that the significant distinction is between gas and liquid, and that in general we must take our chance whether the liquid is oil or water (SITPA 55, 77, 109).
- The dip; if a reservoir is rotated by folding or uplift or subsidence, any gas migrates to the updip end, and so changes the area and thickness of the gas-saturated zone.
- The water saturation, if the hydrocarbon is gas; the only significant distinction, however, is between less-than-a-few-percent of gas and more-than-a-few-percent of gas (SITPA 97–100).
- The cementation, both in the sense that it modifies the original porosity and also in the sense that it is responsible for the velocity dog-leg at shallow depth (as discussed in section 2.2.2 above).
- The depth, in the sense that it modifies the hardness—particularly of contacting shales, rather than of the sand itself—by compaction.

Group 3

This group identifies the factors introduced by the **seismic method itself**.

- The seismic pulse (as the other major component defining the interference pattern between the top and bottom reflections). This pulse depends on the source, the spread and array geometry, the recording filters, the processing—and the earth filter.
- The diameter of the reflection zone, which depends on the pulse, the reflection time, the velocity, and the source-to-geophone distance.

In this chapter we shall be concerned only with Group 1. We ask what are the seismic signatures of sand bodies as decided by the **sediments** and the **depositional environment**. Then we can search for these signatures on our sections—accepting that they are modified by the post-depositional processes of Group 2, and being ready to optimize them by proper attention to the seismic factors of Group 3.

It is not the function of this course to describe in detail the characteristics of the various types of sand body; this has been done admirably in many papers and several courses. However, for summary purposes, we include (as Figure 3–1) the master diagram of LeBlanc and the abstract of his classic paper (1972). Also included is the *Summary of Characteristics of Some Environmental Types of Sand Bodies* given by MacKenzie (1972), reproduced on p. 44. An earlier tabulation by Shelton (1967) is also relevant. The geophysicist requiring to be led by the hand through the verbal maze of the geologist is referred to Chapman's *Petroleum Geology* (1973) and Selley's *Introduction to Sedimentology* (1976); both are eminently readable books. A valuable compilation of case histories is also given by Selley, in his *Ancient Sedimentary Environments* (2nd edition, 1978).

Geologists normally think of sandstone bodies in an orderly progression from continental through shore-line to deep-water units. For our purposes, the exposition is simplified if we adopt a rather different order—basically one of increasing seismic complexity.

3.1 The massive blanket sand

The most likely origin of a very thick sand body is as a sand sea, built up by sand transported by the wind and deposited in dunes; an obvious present-day example is the dunes of the Rub-al-Khali, which represent sand several hundred metres thick.

Such a sand body can be a very good subject for seismic analysis.

- It is fairly homogeneous; we are not troubled by transitional (low-frequency) reflections bounding or within the sand.
- Indeed the bounding reflections may be quite discrete. The lower reflector is likely to be plane;

		GENERAL	CHARACTERISTICS OF ENTIRE SEDIMENT BODY				CHARACTERISTICS OF INDIVIDUAL VERTICAL SECTIONS										
											PRIMARY SEDIMENTARY STRUCTURES						
											CROSS-STRATIFICATION						
		LITHOLOGY	THICKNESS (FEET)	SHAPE, HORIZONTAL DIMENSIONS	DISTRIBUTION, TREND	RELATIONSHIP TO ADJACENT OR ENCLOSING FACIES	LITHOLOGY, COMPOSITION, TEXTURE, FAUNA	BOUNDING CONTACTS	OVERALL VERTICAL GRAIN-SIZE CHANGE	STRATIFICATION	CONTACTS; SET THICKNESSES	NATURE OF LAMINAE	SHAPE OF SETS	RIPPLES	DEFORMATIONAL AND ORGANIC SEDIMENTARY STRUCTURES	MISCELLANEOUS REMARKS	REFERENCES
	SUBAERIAL MIGRATED DUNE SANDS	sands, no muds; homogeneous	10 to 1000	elongate, or sheets up to 1000's of sq. miles in area	downwind from source of sand	commonly the end stage of a regressive sequence	well sorted sands; pebbles & clasts rare	variable	not systematic	conspicuous high-angle cross bedding	erosional, horiz. or sloping. sets up to 10's of feet thick	lee dips 25°-34° commonly tangential to lower boundary	tabular; sometimes enormous troughs	high indices, crests often parallel to dip of lee beds	slumps not uncommon; vertebrate tracks		McKee (1966)
	ALLUVIAL SANDS	sands, muds, some gravels	usually 30-80; sometimes 200-300	continuous bodies; usually ½ to 5 mi. wide, 10's to 100's of miles long	make large angles with shoreline trends	lower contacts erosional; lateral contacts erosional or indeterminate	pebbles and clasts common; proportion of mud variable	base erosional; top usually transitional	upward decrease	many beds lenticular; abundant cross bedding	erosional, planar or concave up; usually ¼-2 feet thick	maximum dips usually 20°-25°, inclined or tangential to lower boundary	trough	short-crested; linguoid; microtrough in cross-section; abundant	slumps common; burrows uncommon		Harms (1966) Hewitt & Morgan (1965) Fisk (1944) Potter (1967)
DELTAIC DEPOSITS	DISTRIBUTARY CHANNEL FILLS	sands, muds	up to 200	continuous sinuous bodies, usually < 1 mi. wide		commonly enclosed in nonmarine or blackish muds									slumps, burrows not uncommon		Frazier (1967) Brown (1969)
DELTAIC DEPOSITS	DELTA-FRONT SHEET SANDS	sands	20-80	sheets		underlain by marine pro-delta muds; overlain by marsh muds	←			similar to barrier sand bodies →							
DELTAIC DEPOSITS	REWORKED TRANSGRESSIVE SANDS	sands	1-40	sheets		underlain by or adjacent to deltaic deposits	well sorted; may contain coarse sand lag	both sharp	not systematic	high-angle cross bedding; orientation diverse	erosional, planar; ¼-2 feet thick	maximum dips 20°-25°; tangential to lower boundary	wedge or tabular	not conspicuous; microtroughs present	slumps, burrows uncommon	lateral facies changes may provide proximity indicators	Frazier (1967) MacKenzie (1965)
	REGRESSIVE SHORELINE SANDS (BARRIER-ISLAND SANDS)	sands; rare muds	20-60	elongate or sheets; up to miles wide, 10's of miles long	parallel to shoreline where elongate	transitional downward and seaward into muds, landward into lagoonal or deltaic deposits	well sorted; pebbles & clasts rare; marine fauna, if any	base transitional; top sharp	upward increase; (but middle may have coarsest beds)	upper & lower: subhorizontal stratification with low-angle truncations, exp. near base; sets < 1 foot thick middle: high-angle cross bedding, ¼-2 feet thick, tangential laminae; cross-laminae dip obliquely shoreward; local scours			middle: wedge or trough	upper & lower: most abundant near base; symm., long- crested middle: uncommon	upper & lower: load structures & burrows common at base middle: uncommon		Bernard et al (1962) Weimer (1966) McCubbin & Brady (1969)
SHALLOW MARINE SANDS	OFFSHORE BARS	sands with mud partings	several to 10's	elliptical lenses, less than a few sq. miles in size	scattered; orientation variable	enclosed in and intertongues laterally with marine muds and silts		sharp, or narrowly transitional	not systematic	low-angle cross bedding	erosional, planar; ~1 ft. thick	most dips < 10°; laminae parallel to lower set boundary	wedge?	common; some symm., long-crested	burrows abundant only in marginal facies	gradual outward decrease in sand/clay may provide proximity indicator	Exum & Harms (1968)
SHALLOW MARINE SANDS	STRIKE-VALLEY SANDS	fine to coarse sands and muds; heterogeneous	10-50	elongate; up to several miles wide, 10's of miles long	parallel to pre-unconformity paleo strike	fills erosional strike valleys; intertongues with marine muds seaward; onlaps landward	pebbles, clasts, glauconite, phosphate, marine fauna	variable		tabular units with high-angle cross bedding dipping parallel to sand body elongation	erosional, planar; ¼-5 feet thick	max. dips 25°-30°; tangential to lower boundary	tabular; sets straight & continuous for 100's of feet	common locally, esp. at toes of x-sets; some are long-crested wave ripples	burrowing common	paleogeologic and paleotopographic maps effective in exploration	McCubbin (1969)
DEEP-WATER SANDS	PROXIMAL TURBIDITES	interbedded sands, silts and muds	100's to 1000's	fans or sheets up to 1000's of sq. miles in area	high flanks of deep basins near sand source	may be middle part of regressive sequence from deep to shallow-water deposits	graded bdg; displaced shallow-water fauna; proximal turbidites often with interbedded debris beds			parallel stratified or structureless; may have large mud-lined scours	trough-shaped sets found rarely			asymmetric ripples, both short and long-crested, found at tops of individual beds	burrows uncommon; bedding plane tracks and trails often present	cf. distal beds, proximal beds are thicker, coarser grained, less well graded, less regular, more deformed, and more porous and permeable	Walker (1966, 1967)
DEEP-WATER SANDS	DISTAL TURBIDITES				sumps of deep basins*	sands interbedded with deep-water muds				1-3' continuous beds; parallel or ripple stratified							

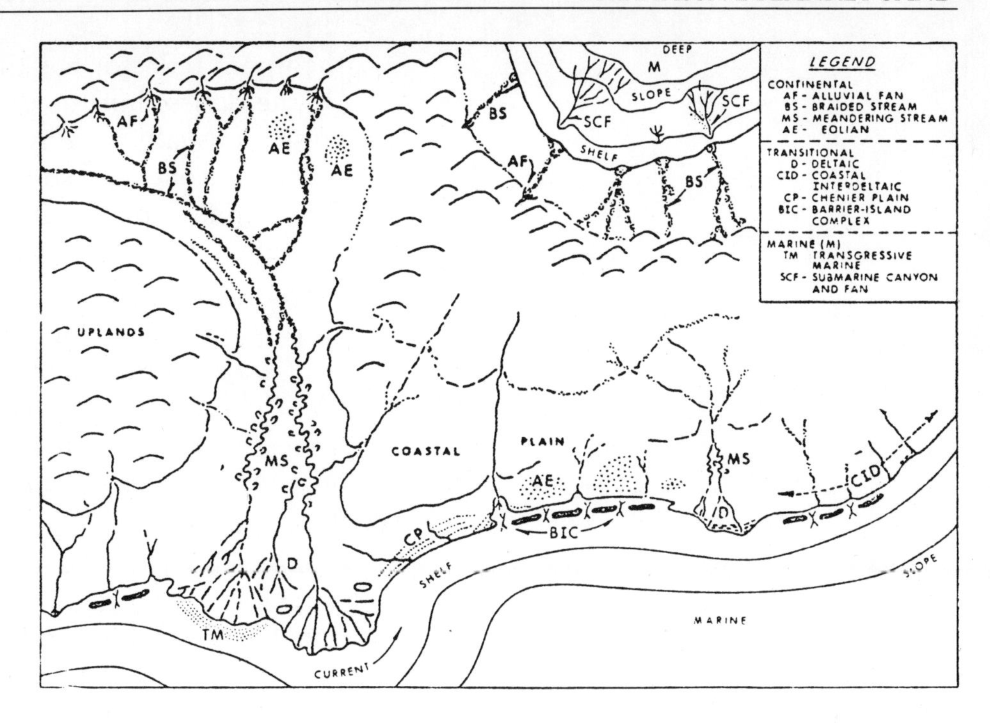

LEBLANC, RUFUS J., SR., Shell Oil Co., Houston, Tex.

Geometry of Sandstone Reservoir Bodies

Natural underground reservoirs capable of containing water, petroleum, and gases consist of sandstone, limestone, dolomite, and fractured rocks of various types. The character and distribution of sandstone and carbonate reservoirs are well known as a result of extensive research and exploration by the petroleum industry. Trends of certain sandstones are predictable because they are more regular and have been less affected than carbonates by postdepositional cementation and compaction.

The principal sandstone-generating environments are (1) fluvial environments such as alluvial fans, braided streams, and meandering streams, (2) distributary channel and delta-front environments of various types of deltas, (3) coastal barrier islands, tidal channels, and chenier plains, (4) desert and coastal eolian plains, and (5) deeper marine environments where the sands are distributed by both normal and density currents.

The alluvial-fan environment is characterized by flash floods and mud or debris flows which deposit the coarsest and most irregular sand bodies. Braided streams have numerous shallow channels separated by broad sand bars. Lateral channel migration results in the deposition of thin, lenticular sand bodies. Meandering streams migrate within belts 20 times their channel widths and deposit two very common types of sands. Bank caving and point-bar accretion processes result in lateral channel migration and the formation of sand bodies (point bars) within each meander loop. Natural cutoffs and channel diversions result in the abandonment of individual meanders and long channel segments respectively. Rapidly abandoned channels are filled with some sand but predominantly with fine-grained sediments (clay plugs), whereas gradually abandoned channels are filled mainly with sands and silts.

The most common sandstone reservoirs are of deltaic origin. They are laterally equivalent to fluvial sands and prodelta and marine clays and consist of two types: delta front or fringe sands and abandoned distributary channel sands. Fringe sands are sheetlike and their landward margins are abrupt (against organic clays of the deltaic plain). Seaward these sands grade into the finer prodelta and marine sediments. Distributary channel sandstones are narrow, with abrupt basal contacts and decrease in grain size upward. They cut into or completely through the fringe sands, and are also connected with the upstream fluvial sands of braided or meandering streams.

Some of the most porous and permeable sandstone reservoirs were deposited in the coastal interdeltaic realm of sedimentation. They consist of well-sorted beach and shoreface sands associated with barrier islands and tidal channels which occur between barriers. Barrier sand bodies are long and narrow, aligned parallel with the coastline, and characterized by a fine to coarse upward sedimentary sequence. They are flanked on the landward side by lagoonal clays and on the opposite side by marine clays. Tidal channel sand bodies have abrupt basal contacts, range in grain size from coarse to fine upward. Laterally they merge with barrier sands and grade into the finer sediments of tidal deltas and mudflats.

The most porous and permeable sandstone reservoirs are products of wind activity in coastal and desert regions. Wind-laid sands are typically very well sorted, highly crossbedded, and occur as extensive sheets.

Marine sandstones are those associated with normal marine processes of the continental shelf, slope, and deep, and those which are of density or turbidite current origin. An important type of normal marine sandstone is formed during marine transgressions. Although these sandstones are very thin, they are very distinctive and widespread, have sharp updip limits and grade seaward into marine shales. Two other types of shallow-marine sands (delta fringe and barrier shoreface) have previously been mentioned.

Many turbidites are associated with submarine canyons. These sands are transported from nearshore environments seaward through canyons and deposited on submarine fans in deep marine basins. Another type of turbidite forms as a result of slumping of deltaic facies occurring at shelf edges. Turbidite sandstones are usually associated with thick marine shales.

FIGURE 3.1 *(After MacKenzie)*

the upper reflector has the roughness represented by the dunes, but the distance between the dune crests is usually smaller than the diameter of the reflection zone, and so the upper reflector appears substantially plane and of about the correct strength.

- If we have sufficient control to know where the sand is on the section, we scan for structural or fault traps in the usual manner. Then we search for a change in the character of the top reflection—or better still a visible fluid contact—at or near the top of the trap.
- If we are lucky enough to have a fluid contact, as in Figure 3.1–1, we set about solving for the porosity of the sand. To do this we need to be confident that the top-sand and fluid-contact reflections are abrupt and discrete, that we have sufficient gas column (say 30-40 ms) to be able to get the velocity in the gas-saturated material, that we can trust the processed amplitudes, and that we can assume the reservoir is a clean wind-winnowed sand. There are three methods available to us for calculating the porosity, and we use whichever one (or ones) fits the situation (SITPA 144-168, 175-176, 337-395).

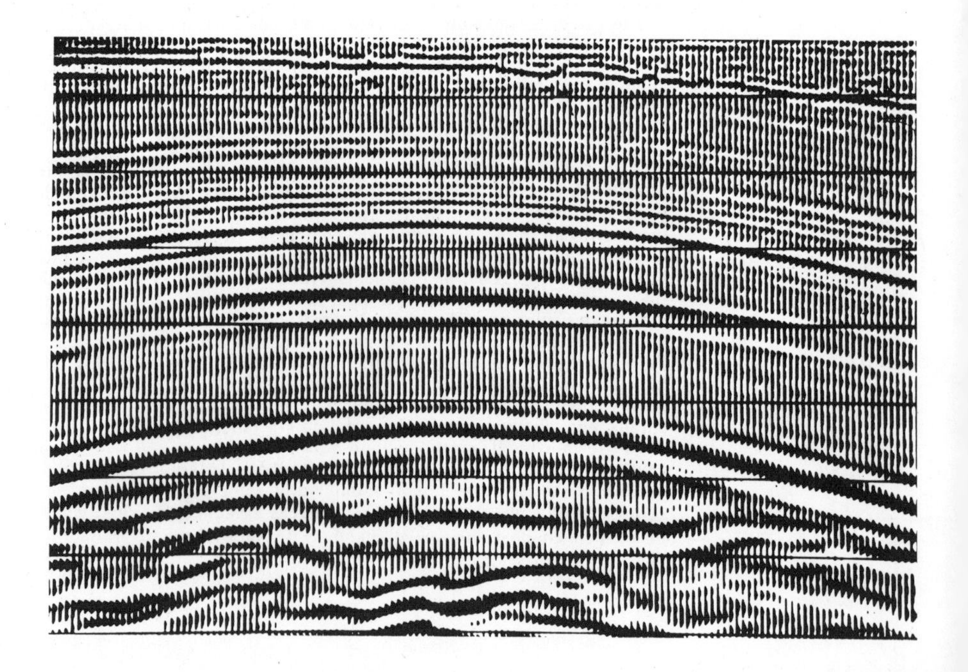

FIGURE 3.1–1

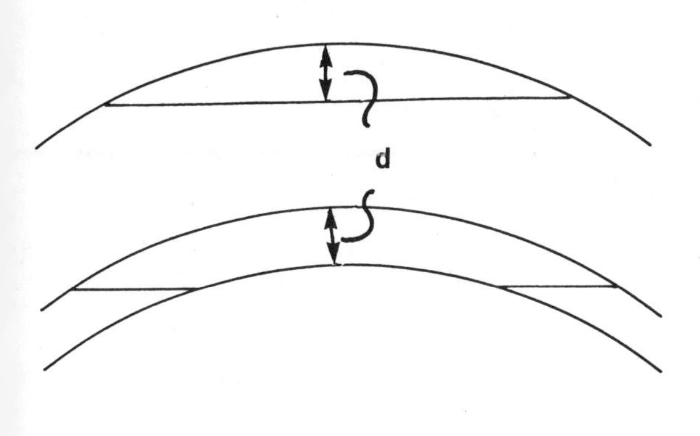

Saturant	d	Reservoir properties (in addition to areal extent) which we can compute under favourable circumstances
GAS	greater than t*	**thickness** (in metres or feet) **acoustic impedance** **density** (for sand reservoir) **porosity** (within bounds of water-saturation)
	between t and t/2	**thickness** in milliseconds (not metres or feet); if amplitudes are calibrated in units of reflection coefficient, also **acoustic impedance**
	less than t/2	if amplitudes are calibrated, **general figure-of-merit** representing blend of porosity and net thickness in milliseconds
OIL	greater than t	approximate **porosity** (for sand reservoir) but no indication whether oil is present
	less than t	nothing

* t is the period of the seismic pulse generally observed on reflections near (but not at) the target level.

FIGURE 3.1–2

- We can do the same without a clear fluid contact if we can see evidence (in a polarity change on the top reflection, for example) that the entire thickness of the sand is fully gas-saturated at the top of the trap (lower sketch of Figure 3.1–2).
- Finally we establish the possible balance between porosity and water-saturation (SITPA 346), and proceed to validate our interpretation by modelling (SITPA 467-479).
- The properties of the reservoir we can hope to resolve, therefore, are those of the "greater than t" section of the table in Figure 3.1–2: thickness, acoustic impedance, density and porosity. Not bad.
- But probably we are not so fortunate, and the gas-saturated thickness of the sand is less than 30-40 ms. Then we look for a character change in the top reflector, in the vicinity where we expect the gas. The picture would be like 3.1–1, but with less gas the fluid-contact reflection would interfere with the top-gas reflection to form one complex.
- If the character change is to a form like the upper traces of the Widess diagram (Figure 2.4–3)—and in particular if it shows the hickies of those waveforms—we first establish the basic pulse shape from the section (SITPA 213-219, 226, 229, 230, 473, 474, 540-546), then construct our own Widess diagram for that pulse, and finally search for a match between these modelled traces and the section. In effect, we are modelling to see what thickness of the gas-saturated interval would give the observed character change. For this, of course, we assume that the top reflection is negative, and the fluid-contact reflection interfering with it is (as it must be) positive.
- It is important that we should not be afraid of this modelling step. If we have an interactive terminal in the office, well and good—it does not even look like work. But if not, we will do it by hand—at least initially—and find it no great effort either.

We will select a nearby reflection which we think (on lithologic considerations, or on considerations of simplicity) to be discrete. We will find a trace which looks average for that reflection, and pencil it on to two pieces of tracing paper. We will add our best estimate of the centre-

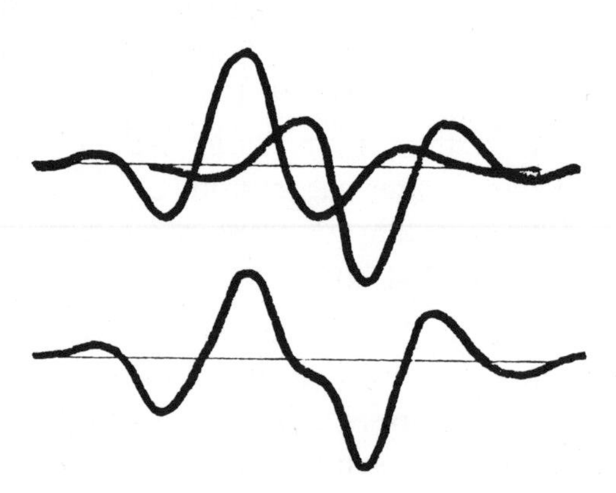

line position of the trace. Then we will turn one piece of paper upside-down (so that one reflection is positive and the other negative—we do not yet know which), and slide the two pulses back and forth along the common centre-line to represent different times between them. In a moment, we will add them together in various time positions, but first we overlay them on our gas complex and try various combinations of time positions for the two possible polarities. This usually establishes the polarity, and narrows the range of time positions. Then, for several values of time shift, we merely add the two pulses together by eye— conveniently, just as peaks, troughs, zero-crossings and trace-crossings—and draw the composite pulse. After the first time, it takes perhaps 10 minutes. We can send it to the computer to be done properly—including changing the relative amplitudes of the two pulses—after we have satisfied ourselves that it is interesting enough to warrant the cost; we may believe the machine's result more than ours, but we certainly will not understand it as well.

- To all this we add whatever knowledge we can glean about the position of the fluid contact from the down-dip character changes of the top-gas reflection. In the case we have assumed above, we expect a polarity change indicating the position of the contact, but of course the amplitude must be vanishingly small at this point. It would be easy to say, "Look for the first change in the top reflection, as we come updip to the trap"—but we must remember the blurring effect of the reflection zone as it approaches and coincides with the edge of the gas. We cannot expect the edge of the gas to be as clear as a pikestaff; as Bortfeld has said, "There are discontinuities in the earth, but there are no discontinuities on seismic sections." Everything is smeared over by the reflection zone.
- For this intermediate thickness of the reservoir, from which we derive a reflection complex with a discernible *character*, the properties of the reservoir we can hope to resolve are those of the section "between t and 1/2t" in figure 3.1-2: thickness (in milliseconds, now) and acoustic impedance.

- Perhaps we have even less gas column, so that no hickies appear in the complex reflection from the gas. In that case, we expect the complex to show the *differentiated* form (higher frequency, skew-symmetric) discussed in section 2.4. Then we can learn nothing about the thickness of the gas column by studying pulse shapes; our only guide is **amplitude.**
- We can calibrate the observed amplitude of the complex by reference to that of a nearby discrete reflection, using the techniques of SITPA 144-168; in effect, this gives us an amplitude-measuring scale calibrated in reflection coefficient. If the observed amplitude of the complex is large, on this absolute scale, then we know **either** that the gas column is thick (up to a limit of 1/4 λ, beyond which we would begin to see hickies in the waveform) **or** that the sand is very porous (Lindsey, Dedman and Ausburn, 1978). More than this we cannot say; in particular we cannot say how porous is *very porous.*
- In this situation we are in the section "less than 1/2t" in figure 3.1-2; all we can obtain is a gross figure-of-merit for the reservoir, representing some measure of its gas reserves. In this context we are perhaps luckier than we deserve; figure 3.1-3 (reproduced from Lindsey, Schramm and Nemeth, 1977) shows that the amplitudes follow the **net** thickness of sand surprisingly well.
- But perhaps we do not have gas at all, but oil. Then what we are hoping for is **weakness** in the reflection from the top of the sand, if it is overlain by a shale. (A strong positive reflection would mean poor porosity; a strong negative reflection, for liquid saturation, could mean only that the overlying material is a carbonate.)
- So the determination of reflection polarity is important. The general absence of a detectable oil-water reflection means that polarity determination is easier with oil than with gas; there is less interference to complicate the pulse shapes. The polarity determination is made by maintaining a properly-controlled recording and processing system (including phase-zeroizing), and/or by the techniques of SITPA 219-223, 268-277, 377, 538.

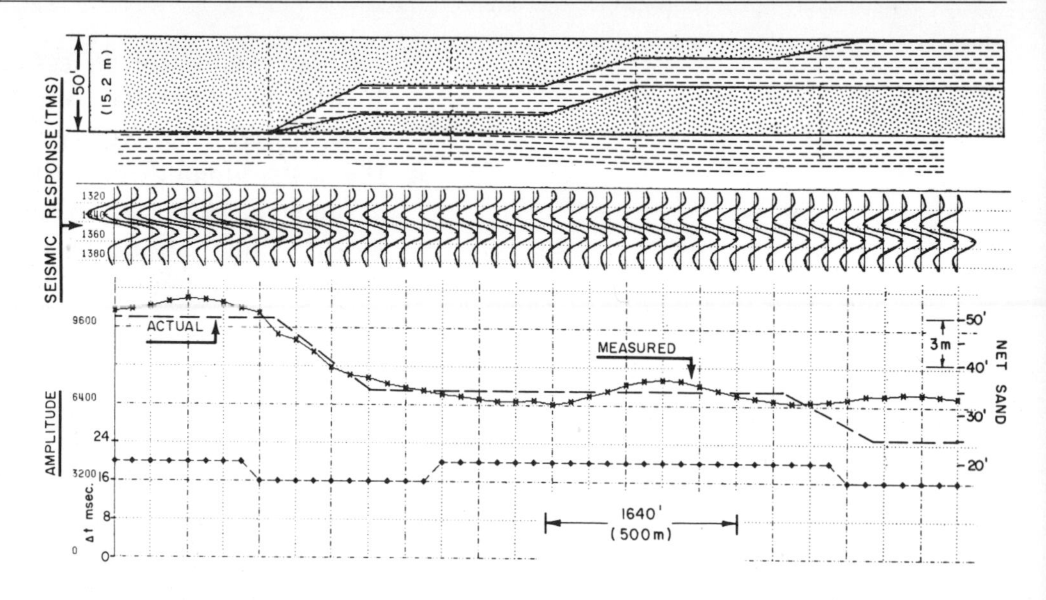

FIGURE 3.1–3
(After Schramm et al., 1977)
Effects of a shale stringer in a thin sandstone on seismic reflection from sandstone body. Seismic wavelet shape does not vary with shale thickness or distribution in the sandstone. Amplitude, however, is directly related to net amount of sandstone.

- If we have weak top-sand amplitudes, and hopes of oil, what can we do to estimate the thickness of the oil column? Nothing—because we do not ordinarily expect any seismic indication of the oil-water contact.*
- But we can calculate the porosity. All we need to do is to determine the interval velocity in the sand (very carefully, using the techniques of SITPA 300-307). To do this we need a sand thickness corresponding to the "greater than t" criterion of figure 3.1–2 (now in the *oil* category). The reason we can find the porosity with **velocity**, instead of having to wrestle with the greater uncertainties associated with amplitudes, is that there is a simple equation connecting porosity and velocity for the case of liquid-saturated sands; this equation does not apply to the case of gas saturation (SITPA 64, 77-79, 94-100). The major proviso concerns the dog-leg of section 2.2.2; if the sand is angular, it must be old enough to be at least slightly cemented. It must also be normally pressured.
- So if we have the optimism to expect that our thick sand is fully saturated with oil, we can calculate the porosity. But when we drill, we may find a little more porosity—full of water. The only hopes of seismic differentiation between oil and

* Unless there are marked differences of porosity in the two zones, brought about by continuing cementation of the water-saturated zone (SITPA 412).

water, as such, lie in the observation of small shallow bright-spots indicating past gas leakage from a deep oil reservoir (SITPA 406-409).

- If the thickness of the oil sand is insufficient to allow a velocity determination (last section of figure 3.1–2) we can deduce **nothing** about the reservoir.

Perhaps we need a summary of this section. Here goes:

- A thick homogeneous sand is unlikely to declare itself positively by any internal characteristics obvious on the seismic section; however, any stratification we see within it should not be precisely conformable.
- In a regional sense, it should be in a position compatible with an origin as wind-blown sand dunes (that is, with the final stage of a regression).
- Such a sand is the ideal subject for quantitative seismic analysis.
- If the sand is overlain by shale, a strong positive reflection from its top means that the sand is liquid-saturated and tight. A weak reflection means that the sand is **either** liquid-saturated and porous **or** gas-saturated and tight. A strong negative reflection means that the sand is gas-saturated and porous.
- If the sand is overlain by a carbonate, the reflection is negative. A fair reflection indicates the sand is liquid-saturated and tight, a strong reflection indicates it is either liquid-saturated and porous or gas-saturated and tight, and the grand-daddy of all strong reflections indicates it is gas-saturated and porous.
- If the sand is liquid-saturated, we can calculate its porosity from its velocity. The value obtained should be compatible with the amplitude indications as above.
- We cannot distinguish oil from water, in general.

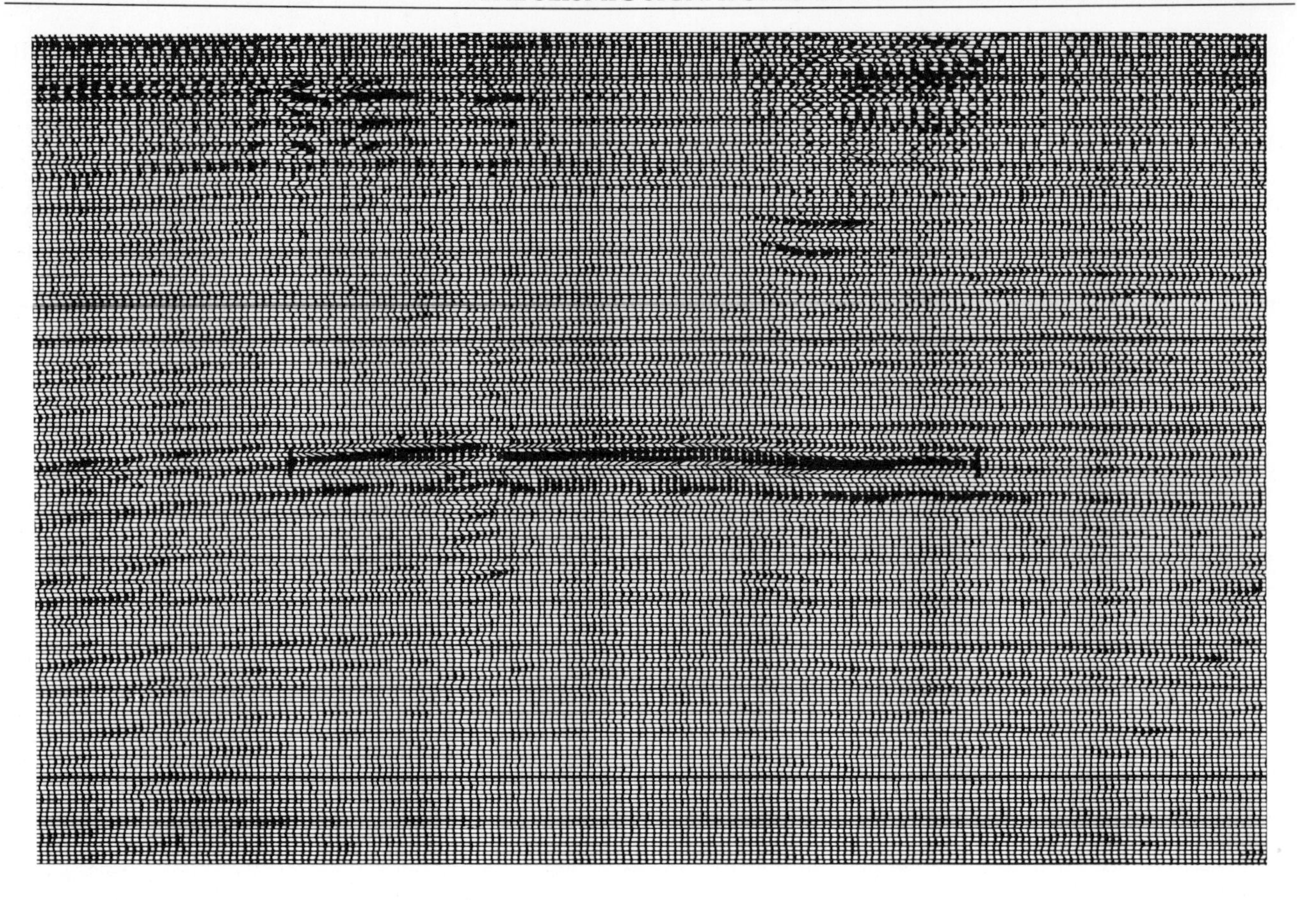

FIGURE 3.1–4 Exercise: *How would we use the Widess diagram to establish the thickness of this sand?*

- If the sand is gas-saturated, we look for a fluid contact.
- If the fluid contact is clear we can calculate the porosity, through density, using both amplitudes and velocities. Then we also know the thickness of the gas column, and so can estimate the reserves.
- If the fluid contact interferes visibly with the top-sand reflection, we can hope to obtain the time thickness—but not the depth thickness—by modelling, and some gross feeling for the porosity from the amplitudes.
- If the fluid contact blends smoothly with the top-sand reflection, we cannot separate porosity and thickness; we can say only, "the stronger the better."

3.2 The thin blanket sandstone

Thin, in a seismic sense, means of such thickness that the top reflection and the bottom reflection interfere. This criterion includes the actual thickness, the velocity, and the form of the seismic pulse. According to the best-case conditions we discussed in connection with figure 2.4-1, a bed starts to become thin somewhere around 40 m (130 ft). Deeper in the section it would be more than this.

It is in the nature of sandstones that most of them are thin, in this sense—even under our best-case conditions.

Several depositional mechanisms can lead to thin blanket sands. The major categories are delta-front sheet sands, transgressive shoreline sands and regressive shoreline sands. Thicknesses for such units range up to 25 m (80 ft). The dune-sand mechanism of the last section can also yield thicknesses in this range.

When we talk of blanket or sheet sands, in this section and the last, we imply that the major recognition criterion is not based on the **areal** configuration of the sand body. Recognition depends first on what we see on the section—in the vertical plane.

Let us see what are the tools for this recognition.

3.2.1. The transgressive shoreline sand

Let us start with the hypothetical situation of figure 3.2.1-1a. In this the relative sea level (rsl) is rising; the supply of sediments, while abundant, is not sufficient to prevent a transgression. A thin but continuous blanket of sandstone is deposited unconformably on the original inclined land surface (with the possibility of some non-marine deposits between the two). The sandstone merges transitionally with overlying shale. The shale is recognizable seismically by a weak grain of reflections (SITPA 417-419). The geological time lines cross the sand body at an acute angle, as suggested in the figure.

If we make a seismic observation at the location indicated, we encounter a vertical variation of acoustic impedance which is likely to be one of those shown at the right. The first two show the transition from shale into liquid-saturated sand of medium-to-poor porosity; the third shows the transition from shale into gas-saturated sand of good porosity. Below these transitions

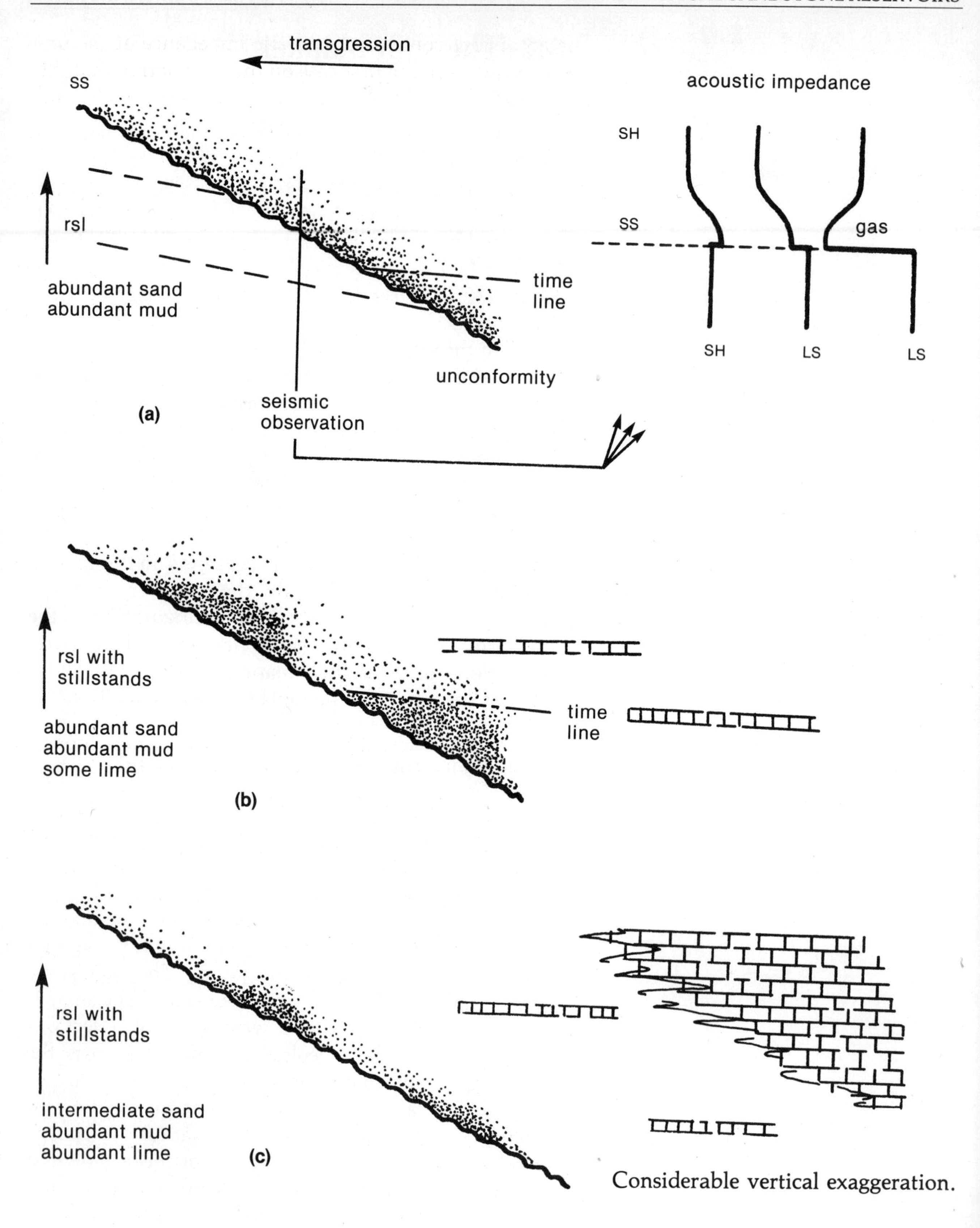

FIGURE 3.2.1–1 *(Generally after Busch, 1974)*

is an abrupt contrast of acoustic impedance at the unconformity—in the first case on the assumption that the material below the unconformity is an older shale, and in the others that it is a carbonate.

The reflection complex that we would observe on the trace, in these situations, depends on the thickness of the sand and the thickness of the shale-sand transition zone. If the sand thickness is 15 m (50 ft) and the transition thickness is the same, then we would not see the distinction between this and a simple thin bed; the pulse shape would be differentiated and the amplitude would depend on the sand thickness. But if the transition occurs over 75 m (250 ft), the effect is quite different; we get only a weak low-frequency reflection from the transition, and then a clear discrete reflection, of normal shape, from the unconformity. Thus in the first and third cases at the right of the figure, in which the top-sand reflection and the base-sand reflection are of opposite polarity, the presence of the slow transition **increases** the obviousness of the sand; the amplitude of the complex depends much less on the thickness of the sand. The reflection form we expect is similar to that of sketch 12 in figure 2.5-1.

We note the catch: the visually-obvious reflection is the one from the **base** of the reservoir—which is **positive** for a porous gas-saturated sand.

Further, we must always remember the equal importance of **both** materials across a reflection-generating interface. Thus in the sketch at the left we suggest by dashed lines that the original land mass itself contains dipping layers truncated at the unconformity; now let us suppose that within the dashed lines is a carbonate, while above and below are shales. Then the reflection we see from the base of the sand, if the latter is liquid-saturated, flips between the first and second conditions on the right—it flips polarity. Thus the presence of reflections truncated at the unconformity would be our warning not to take the message of the base-sand reflection at its face value.

The seismic signature of the hypothetical sand we have described is therefore the basic thin-bed response, superimposed on and modifying the basic unconformity reflection which would occur in the absence of the sand. However, if the top of the sand is transitional into shale over a considerable thickness, we might see no more than a suggestion of a low-frequency wiggle just before the basic unconformity reflection. If the sand is very thin it scarcely appears at all; we obtain the reflection we would have obtained from the unconformity surface if the sand had not been there.

We have stressed that this situation is hypothetical. Is it geologically possible to have a transgressive sand sheet of substantially uniform thickness lying along the unconformity, at an angle to its bedding planes? It may be rare, but we cannot say it is impossible; the appropriate conditions of sediment supply and uniform change of relative sea level cannot be ruled out. Let us say that where it occurs it cannot be expected to maintain a uniform sheet over very large distances. Do we see the seismic signature on real sections? Probably we do not know. If the upwards transition into shale were rapid it would be difficult to distinguish the seismic response of the combination of sand and unconformity from the characteristically variable signature of the unconformity alone. If the transition were slow, prior practice would probably have removed the low-frequency wiggle in the filters, or delayed it so that its nature as a precursor was no longer obvious.

We might also add that a uniform sand sheet of the type hypothesized would be unlikely to make a good petroleum reservoir; the chances are that the pores would contain significant clay.

Much more general, and much more attractive as a petroleum reservoir, is the situation of figure 3.2.1–1b. Here we recognize that significant deposition of sand in a marine transgression is usually restricted to periods of stillstand. The sand accumulations prograde seaward, and their edges may be sharpened and winnowed by wave action to form terraces of clean well-sorted sand. The seismic signature becomes that of multiple thin wedges. Because the zones of thickness are local, the observed reflections are local also. To the extent that the rise in sea level between stillstands may be expected to be fairly rapid, low-frequency precursors are less likely. Further, the reflections as observed correspond much more nearly to time-stratigraphic surfaces; if the wedge is thick the top-sand reflection (which is time-stratigraphic) may be seen, while if it is not the resulting inteference complex may at least be seen to be inclined to the unconformity surface.

An additional characteristic—limestone stringers—are also shown in the sketch. Such thin limestones may be deposited offshore during the stillstands; these are likely to show as weak local reflections (weak only because of their thinness) whose attitudes and locations confirm the conclusion of a transgressive situation.

The third sketch shows the effect of having less sand and more lime. The sand is still sufficient to give a continuous sheet (though of local thickness during stillstands, of course); we reserve to a later discussion the discontinuous shoreline sands. The sand itself is more difficult to detect seismically—but some help comes from the limestone. The presence of the massive limestone (detectable as such from its interval velocity) is a corroborative indication of the transgression if its base generates a transitional low-frequency reflection; we are then encouraged to look for our transgressive blanket sand at the unconformity below the underlying shale. The limestone stringers may even **point** toward the thicker parts of the sand.

The end of the transgressive sequence may be marked by the beginning of a regression (discussed next), or by a rapid fall in relative sea level (SITPA 435-437). In the latter case the upper boundary of the sequence is likely to be an unconformity showing erosional topography.

From the discussion of all the various circumstances above, let us identify one single principle:

We hope for transgressive shoreline sands whenever we see a shale grain rising at an acute angle to an unconformity, with a reflection complex of local wedge appearance—or at least of variable amplitude and character—marking the sand at the unconformity.

It is only a hope, of course; there may have been no sand in the sediments.

3.2.2 The regressive shoreline sand

Figure 3.2.2–1a illustrates a regressive counterpart to the transgressive case we just considered; we see a rising relative sea level with a superabundance of sediments.

As we can see from the vertical variation of acoustic impedance at the right, the seismic response to this situation should be distinctive. The important reflection-generating interfaces are from non-marine deposits to sandstone (an interface now not unconformable, and probably only slightly transitional), and sandstone to shale (conformable and certainly transitional). Also shown is the interface from shale to the

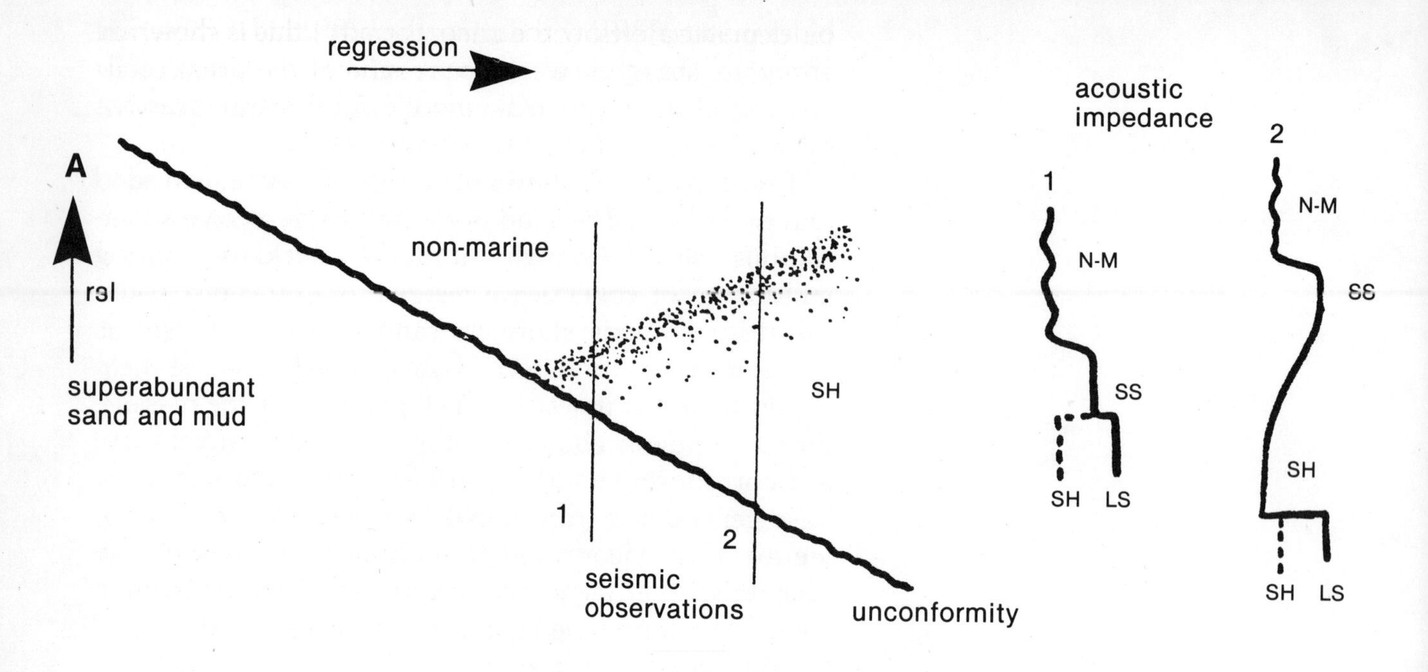

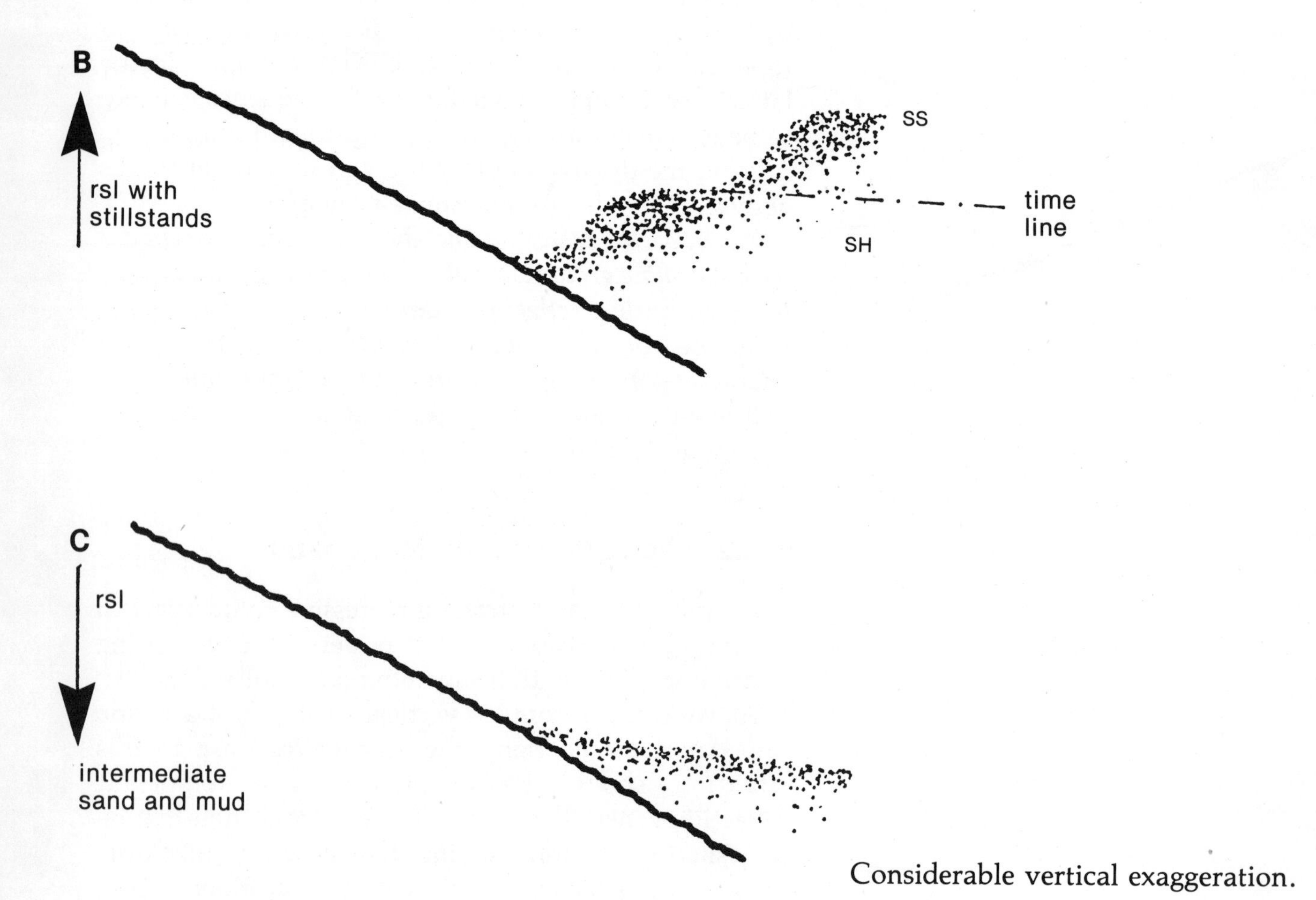

FIGURE 3.2.2–1 *(Generally after Busch, 1974)*

older material below the unconformity; this is shown as abrupt, although in a real expression of the situation illustrated it is likely to be modified by a transgressive sand.

If the transition at the base of the sand occurs quickly, the reflection complex from the sand (as seen on a single trace) is that of the standard thin bed—a differentiated shape, and an amplitude dependent on thickness. There is a possibility, however, that the transition into shale at the base of the sand could be distributed over several tens of metres; in this case the reflection complex would consist of a **top** reflection of substantially normal appearance, followed by a low-frequency **tail**. The obvious top reflection would be generally positive for a liquid-saturated sand of medium-to-poor porosity, and negative for a gas-saturated sand of good porosity. We note the distinction from the transgressive case.

An even clearer distinction, of course, is the fact that the low-amplitude grain characteristic of a marine shale is now seen **below** the sand. The non-marine deposits above the sand are likely to show discontinuous and segmented reflections of variable amplitude (SITPA 421). The unconformity, if present as in the sketch, is likely to show the variations discussed in the last section.

Again, we have to note that the situation of the sketch requires an unusually smooth balance of sediment and rising sea level—a balance which is unlikely to be maintained over large distances and a balance which may be adverse to good reservoir properties. However, the possibility does remain, and we note its implication of a seismic reflection complex which does not represent a time-stratigraphic horizon.

Much more likely, as we discussed before, is that the balance is modified by stillstands; then local thickening of the sand occurs (sketch b), together with improved reservoir properties. We shall hope to see the thickening as a local amplitude change (in the appropriate geological position); perhaps we shall even see some suggestion of the convex nature of the thickened zones. Again everything we see is likely to be a reflection complex, and as such the association of an event with a time-stratigraphic horizon is less than safe—but safer than before.

Although a regressive sand can be formed during a fall in relative sea level (figure 3.2.2-1c), it is normally

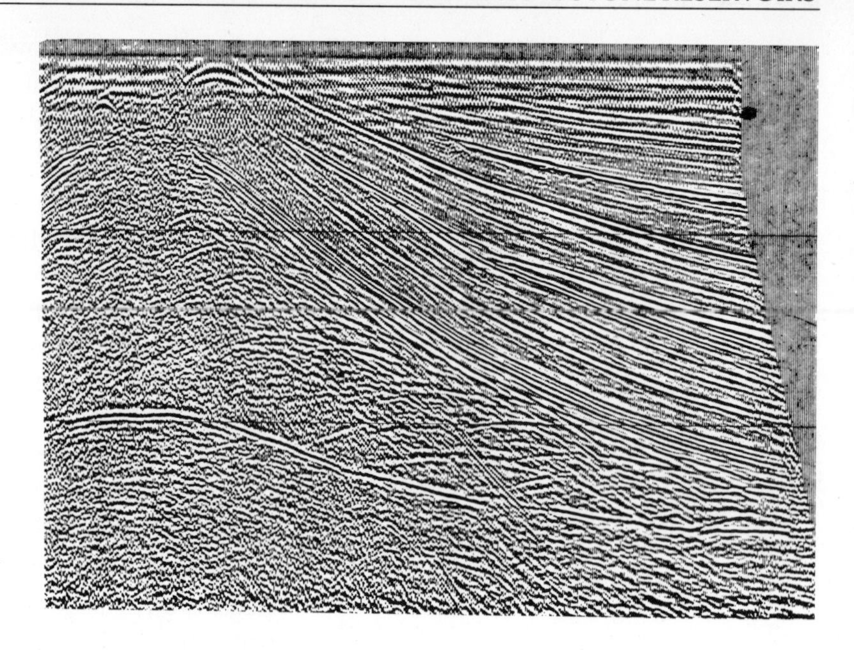

FIGURE 3.2.2–2
Exercise: *Comment on the possibilities for regressive (and transgressive) sheet sands here. (Courtesy Prakla-Seismos)*

eroded during the lowstand and redistributed at the next transgression. This situation is usually too small-scale and too complicated for recognition seismically.

So we are left with a basic signature for only one case, but a very important one—the case of regressive situations with a rising sea level and a superabundance of sediments. The recognition criteria are:

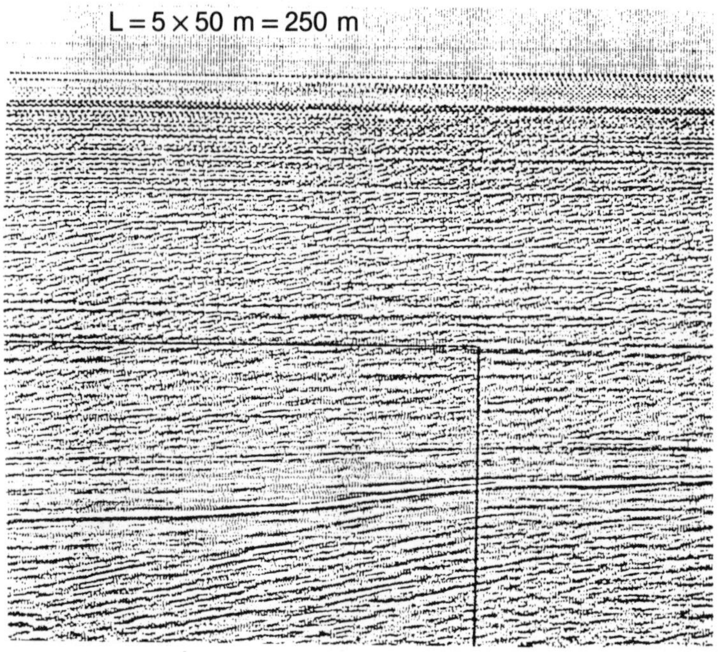

L = 5 × 50 m = 250 m

FIGURE 3.2.2–3 *(Courtesy Amoco)*

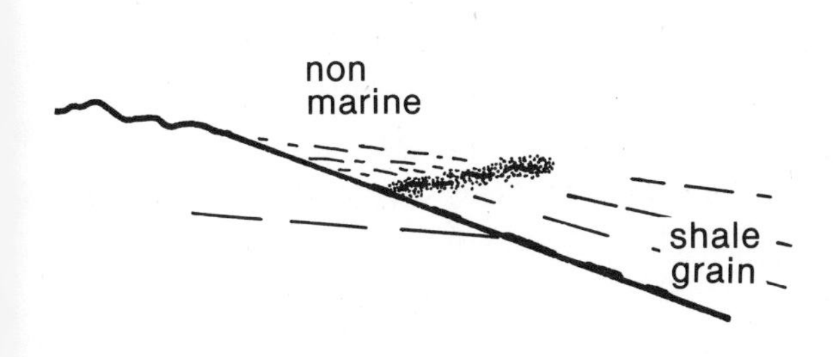

- First, if we have sufficient seismic coverage, we hope to see the relics of a positive sediment source which had sufficient relief to account for the existence and transport of a great mass of material. A possible situation is suggested in the sketch.
- Second, the reflection complex from the sand may be that of a simple thin bed—normally with some changes of thickness expressed as changes of amplitude—or it may show a suggestion of a low-frequency tail associated with a slow transition into the underlying shale.
- Third, the top component of the reflection complex may show a convex appearance suggesting stillstands; in a thinner sand this may show merely as an increase of amplitude, but again in a position compatible with a stillstand interpretation.
- Fourth, the sand reflection complex should represent the demarcation between the continuous but weak reflection grain of the marine shale below and the discontinuous and variable appearance of the continental deposits above.

We cannot pretend that these criteria are always easily recognizable on real sections—particularly when subsequent tectonics are superimposed. However, there is no other course; we must start by considering situational simplifications one at a time, and then hope that we can still recognize them when they are all jumbled together on a real section.

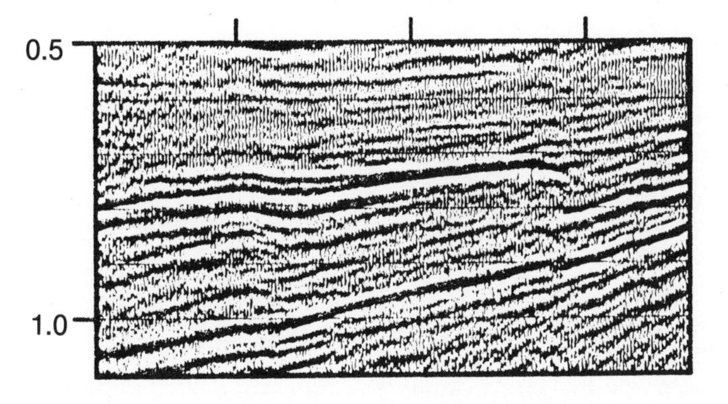

FIGURE 3.2.2–4
(Courtesy S&A Geophysical)

Exercise: *Probable locations for shoreline sands? Of which type?*

3.2.3 The delta-front sheet sand

Two cheers for delta-front sands, for they are often very easily recognized on real sections.

The formation of a delta, as is well-known, requires a superabundance of sediment and a relative sea level which is stationary. As the sediment-laden water discharges into the sea, the coarser sediments are deposited at the top of the foresets, with a steady diminution of grain size down the slope. As the delta builds outward, a near-continuous sheet of silt and sand, locally up to 30 m (100 ft) thick, may be deposited.

The important characteristics of this situation, from the seismic point of view, are that:

- The top-sand reflector is sharp; the base-sand reflector is always transitional.
- Any irregularities in the stillstand situation, or in the sediment load, or in the temperature of the river water during spring flood, leave minor breaks in the continuity of the foreset beds. When these are sufficient to generate reflections, the time-stratigraphic history of the delta is preserved on the seismic section—as in figure 3.2.3–1.

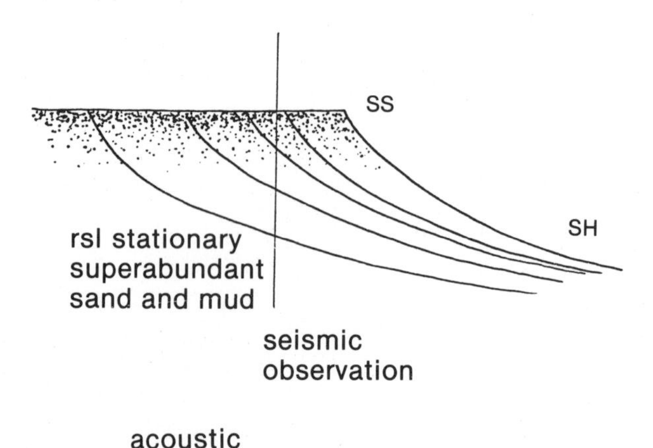

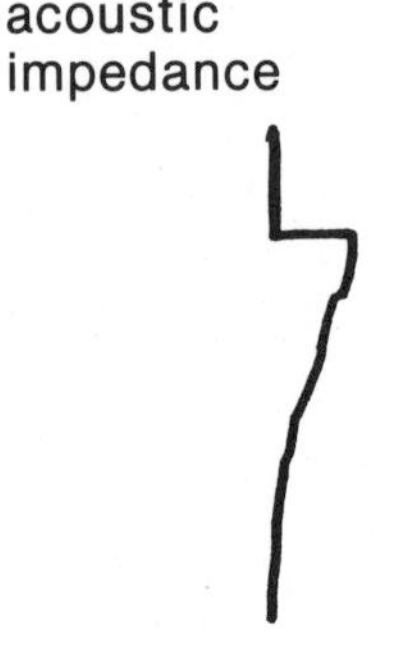

FIGURE 3.2.3–1
Considerable vertical exaggeration

The seismic criterion for recognition of delta-front sands is therefore the progradational appearance of the foreset reflections plus the oblique angle they make with the topset reflection. Further, the topset reflection may have the low-frequency tail seen in sketch 9 of figure 2.5–1 (but with opposite polarity, of course).

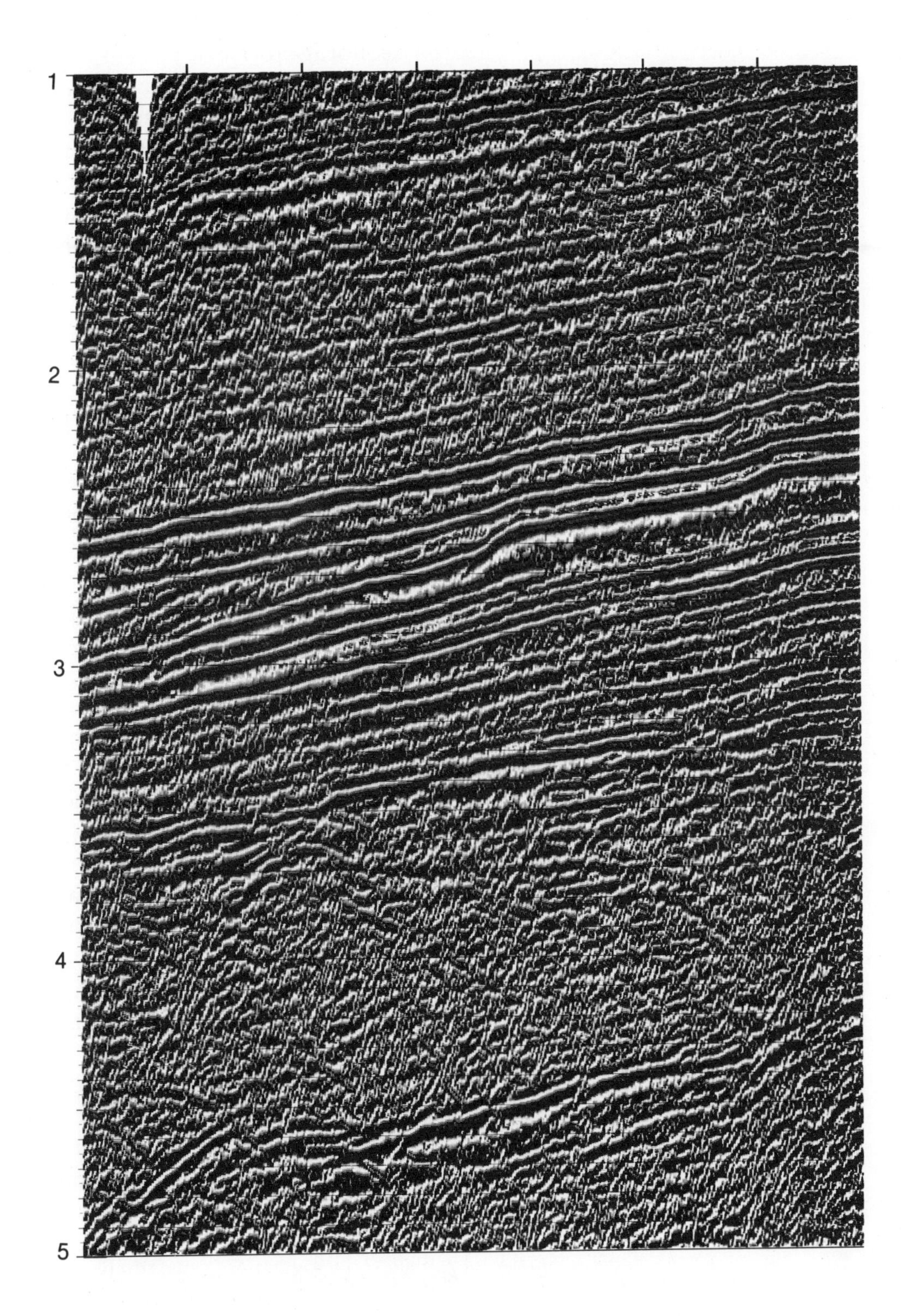

FIGURE 3.2.3–2
Exercise: *Identify at least two locations where delta-front sheet sands are likely. (Courtesy Seiscom Delta)*

As geophysicists, we love to point to the dramatic examples of deltas on seismic sections—the ones with extreme foreset dips. This makes our geological colleagues nervous, because modern deltas do not evince such dips. They are quick to tell us, too, that the deltas which give the better reservoir sands are the ones showing very gradual slopes; the clear seismic expressions are good to look at, but it is the more subtle ones which attract our exploration effort.

We have said, in introducing the topic of blanket sands, that the recognition criteria do not include the **areal** disposition of the sand body. However, prograding delta fronts, and the sands associated with them, often gives rise to some sort of bulge in the shoreline. So if we have sufficient seismic data to add this criterion—and that may mean 1000 km^2—it is fair to do so.

3.3 Unconformity traps

The great appeal of unconformity traps is their possible areal size; we think of East Texas and Prudhoe Bay.

There appears to be no signature feature for a sandstone trap below an unconformity—no way to distinguish the bed as a sandstone, unless some signature feature exists elsewhere along the unit.

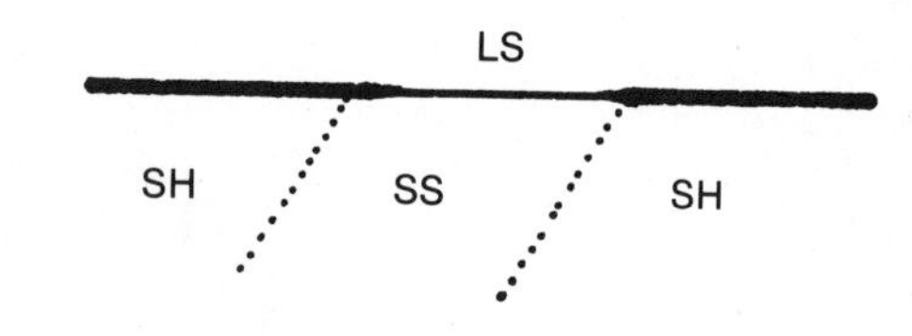

Of course, if we knew we had a sand-shale sequence truncated below an overlying carbonate, and we could see the characteristic strong-weak-strong-weak variation from the unconformity reflection, then we would hope that the zones of weakness represent the sandstone contacts. With an overlying shale, the only cases which would be clear would be the hard tight sand (positive) and the porous gas sand (negative). But these indicators, by themselves, are not as characteristic as the depositional indicators.

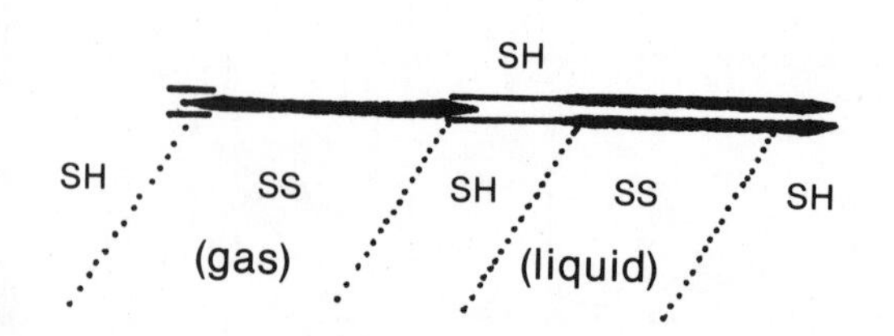

In section 2.7, we mentioned the problems introduced by refraction at an unconformity (SITPA 190, 449, 452, 457). One of these problems is that fluid-contact reflections from just below an unconformity may be grossly displaced on the seismic section. Another is that the reflections bounding the sand may themselves be displaced. This latter problem is serious, because it can lead to the wrong drilling location. So we should explore it further.

Figure 3.3–1, reproduced from SITPA 452, shows a depth model of an unconformity, and its corresponding time section. The question now is: Given the time section, and given one of the formations as the sand target, where do we drill to hit the sand?

Presumably we choose the soft (low-velocity) formation as the target. Then we can see from the figure that:

- The polarity of the unconformity reflection itself is the best indicator of where to drill.
- The reflections from the top and base of the target formation are discontinuous, and they do not reach the unconformity reflection.

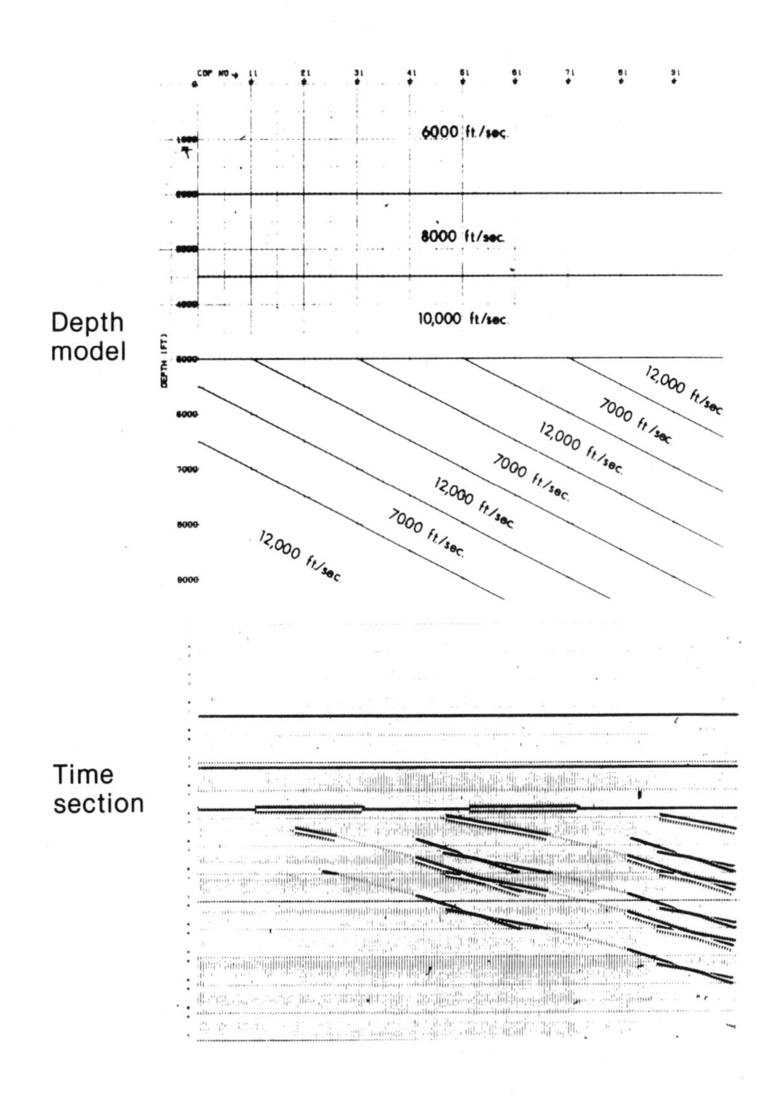

FIGURE 3.3–1

- Extending them upwards to intersect the unconformity leads to a quite erroneous drilling location.
- This is not just a matter of migration. Of course migration is desirable, to correct the normal displacement associated with dip, but it is a formidable problem, in practice, to compensate the additional displacement associated with local refraction.
- So the polarity of the unconformity reflection itself remains the most useful indicator.
- In cases where the material above the unconformity is either harder or softer than **all** the materials below, the unconformity reflection does not flip polarity, and this indicator is lost. Fortunately, in this case, the displacements introduced by refraction are less serious.
- However, even in this milder situation, it cannot be generally correct to search for the limits of a truncated formation without taking account of polarity in some way. Let us say that, as often happens, one of the boundaries of the reservoir formation is positive and one is negative. Let us also assume that the unconformity reflection is negative. Then it would be correct to pick troughs for the unconformity reflection and for the negative boundary of the reservoir bed, and we would hope that (on a migrated section) the intersection of the two trough alignments would be a fair indicator of the position of the truncated boundary. But if we also force a pick on a trough of the other boundary reflection, the intersection of this trough with the unconformity trough emphatically **fails** to indicate the position of that truncated boundary (figure 3.3–2). So we must be prepared to pick the extremum associated with the maximum amplitude—whether it is a peak or a trough.
- Perhaps these points explain a dry hole or two.

Finally we should remember, from section 2-7, that only some of the strength variations we observe at an unconformity are due to genuine changes of acoustic impedance across the unconformity; thinning-bed in-

FIGURE 3.3–2

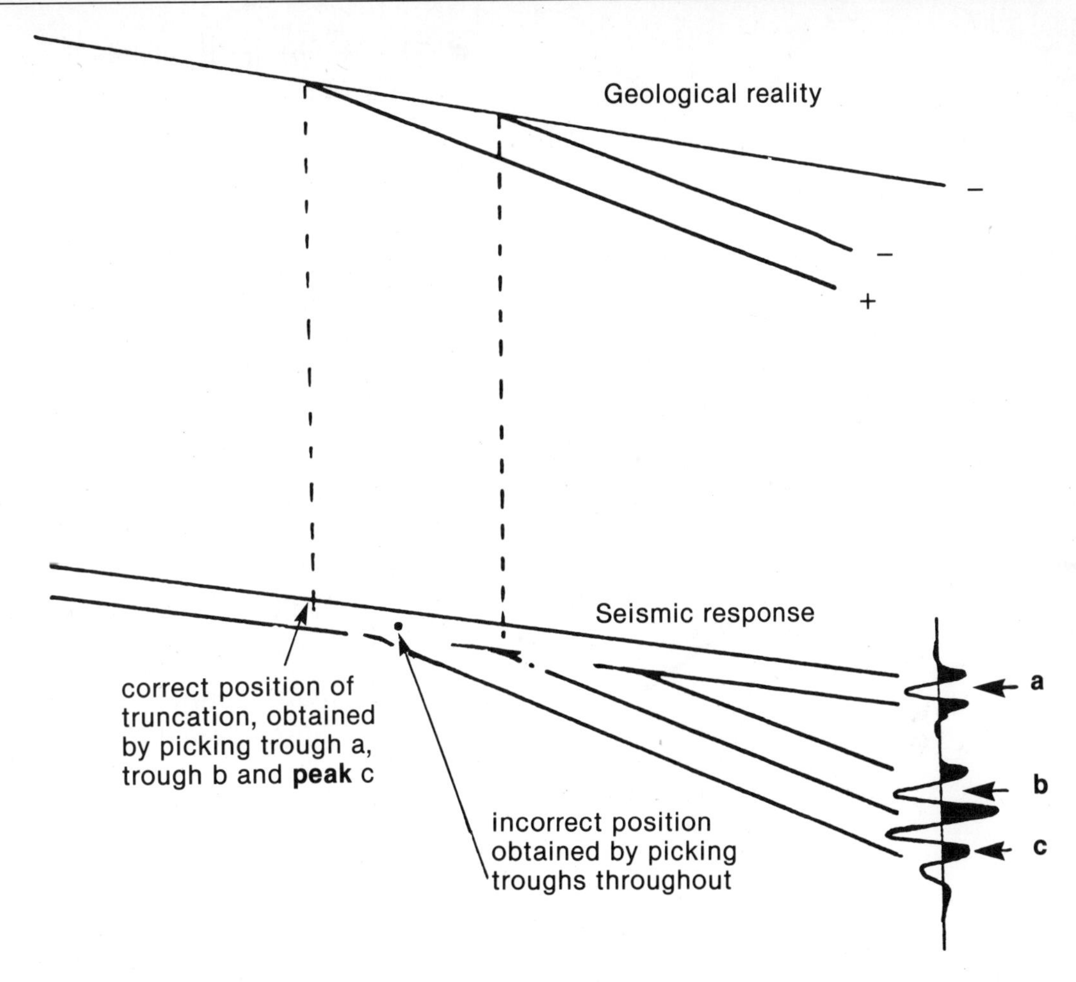

terference itself is sufficient to give a strong-weak-strong appearance along the unconformity. So the important polarity and strength indications are the ones where the truncated and unconformity reflections are separated sufficiently to avoid interference between them.

3.4 Growth-fault reservoirs

We can identify several senses in which growth faults are important in the context of sandstone reservoirs.

(1) Individual growth faults are typically a few kilometres to a few tens of kilometres in length, are concave seaward in plan, are generally parallel to the ancient shoreline, and suggest a significant source of sediment landward of their central region. The last two features, obviously, mean that a growth fault can have the indirect value of **pointing** to sand-prone depositional environments to landward—beaches, deltas and the rest.

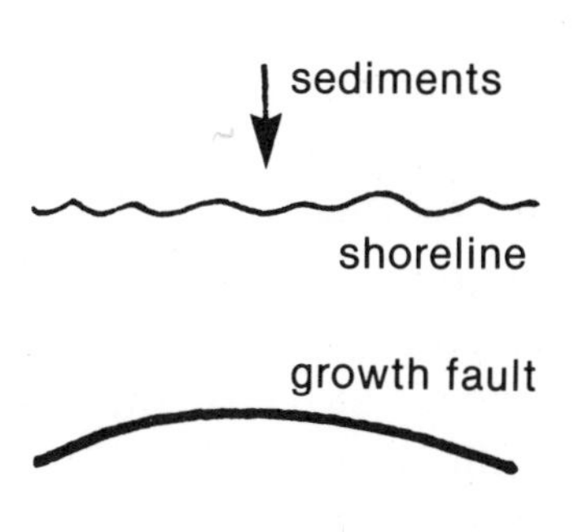

(2) The prime criteria for recognizing a growth fault in seismic section are basically simple:

- The growth itself, visible (in principle) as a thickening of correlatable intervals on the downthrown side.
- The concavity toward the basin in plan, noted above.
- Concavity toward the basin also in section, with angles approaching 60° at the top and attenuating to bedding-plane dip at the bottom.
- In many provinces, a fault-plane merging into the seaward flank of an overpressured shale mass characterized by a reflection-free (or partly chaotic) appearance, poorly defined limits, and very low velocities.
- Often, rollover of the downthrown reflections into the fault, and sometimes a rising of the upthrown reflections towards the fault.

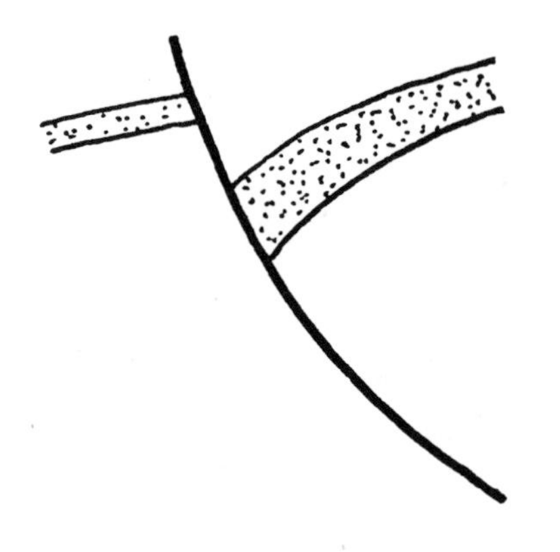

(3) Growth faults cannot form unless sediments are available to allow the downthrown members to be thicker; thus a measure of bottom transport of sediments is essential, and any member showing thickening across the fault cannot have been deposited from suspension. There is therefore a good chance that any such member contains at least some sand. The local thickening of a member containing at least some sand is obviously important in a reservoir sense.

(4) Further, the initiating and sustaining mechanism for a growth fault requires a local loading by **heavy** sediments; the first candidate is sand. The lithologic message of a growth fault is therefore thick sands on the downthrown side. However, since all the bottom-transported sediments swept over the edge contribute to the downthrown thickening, the sands are unlikely to be clean or well-sorted.

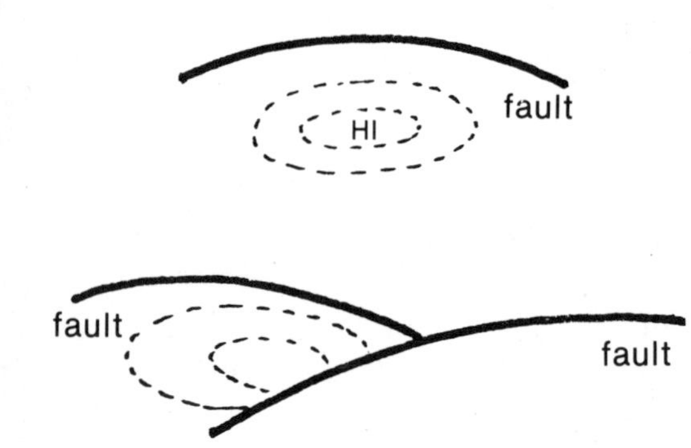

(5) The development of rollover structure on the downthrown side, in conjunction with the concavity in plan, leads to elliptical structures which offer prime and simple targets for seismic exploration. More complex, but often resolvable seismically, is the situation where another fault contributes to the trap.

(6) The crestal axis of rollover structures normally moves basinward at depth, so that a vertical borehole may not give favourable penetration of the multiple pay-zones present. This is illustrated in figure 3.4-1a, from Weber and Daukoru (1975). Also shown are some

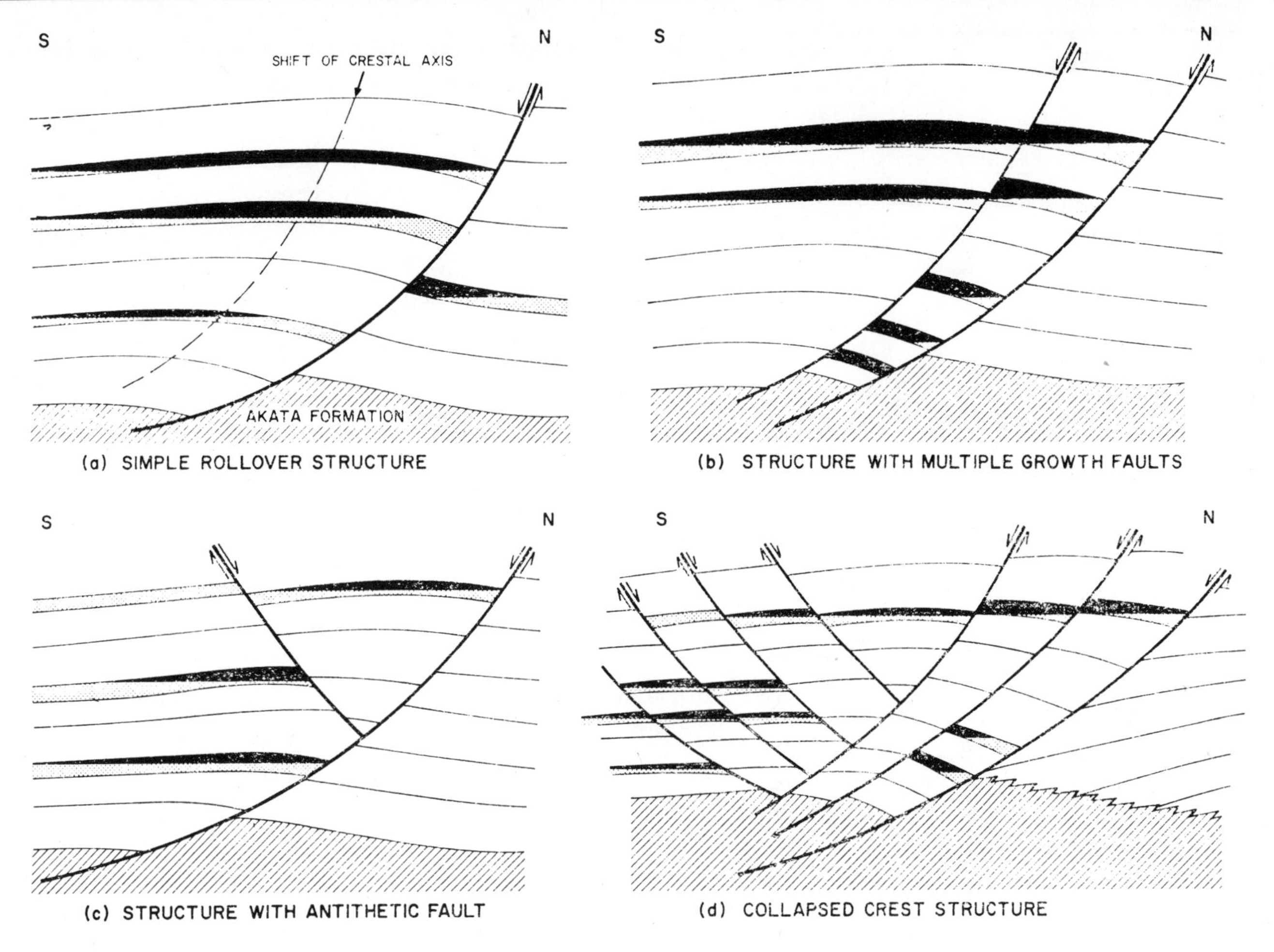

FIGURE 3.4–1 *(From Weber and Daukoru, 1975)*

of the complex fault patterns which can arise; these complicate the interpretation, but give improved possibilities for multiple pay-zones.

With such inducement, we are prepared to slave a little over the interpretation. The mechanism for generation of the main growth fault and the minor associated faults is shown in figure 3.4–2, from a classic paper by Bruce (1973). A seismic section broadly corresponding to the final stage is given in figure 3.4–3.

We can see immediately that the very fact of the downthrown thickening (for example, in sequence 2) is often sufficient to obscure the correlations which would prove the thickening; coupled with rollover and additional faulting, the correlations can become very difficult indeed. (But, even then, perhaps not quite as difficult as on logs.)

We also see that one fault, like one good deed, begets another. In addition to the original growth fault, we also recognize crestal faults and antithetic (or adjustment) faults. Crestal faults, like the original growth fault, are down-to-the-basin; they occur on the crestal portion of the rollover structure, and some tend to merge with the main growth fault at depth. Antithetic faults have the opposite inclination and the opposite throw; they terminate at the original growth fault or at one of the crestal faults. Figure 3.4-4 is a section from Bruce's paper, annotated by Busch (1978) to indicate the original growth faults (g), the crestal faults (c) and the antithetic faults (a). A more detailed classification is given by Evamy, *et al.*,(1978).

We can apply these classifications to identify the mechanism responsible for the growth faulting. In figure 3.4-5 it is clear that each of the two fault systems is associated with a separate shale mass, but that the character of the two systems is distinct. The left-hand system is distinguished by a preponderance of down-to-the-basin normal faults, and is believed to result from differential compaction—simple shear failure in

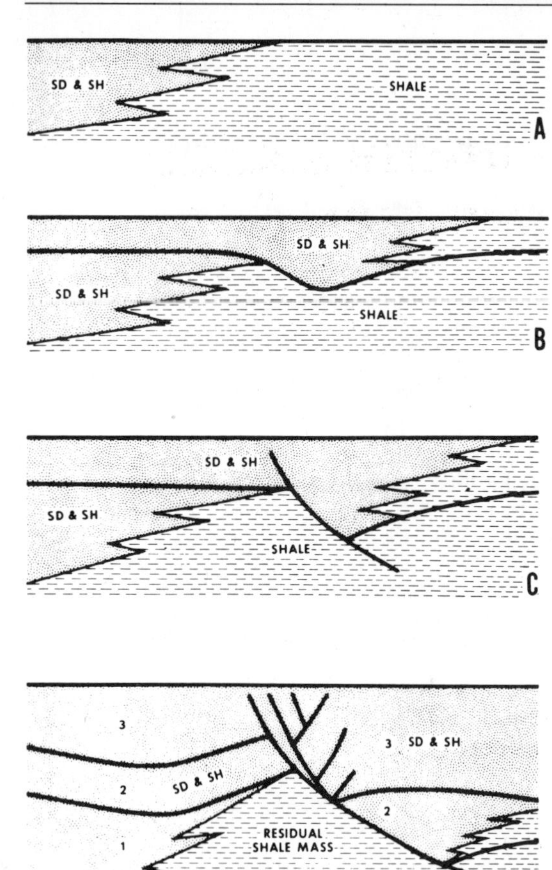

FIGURE 3.4-2

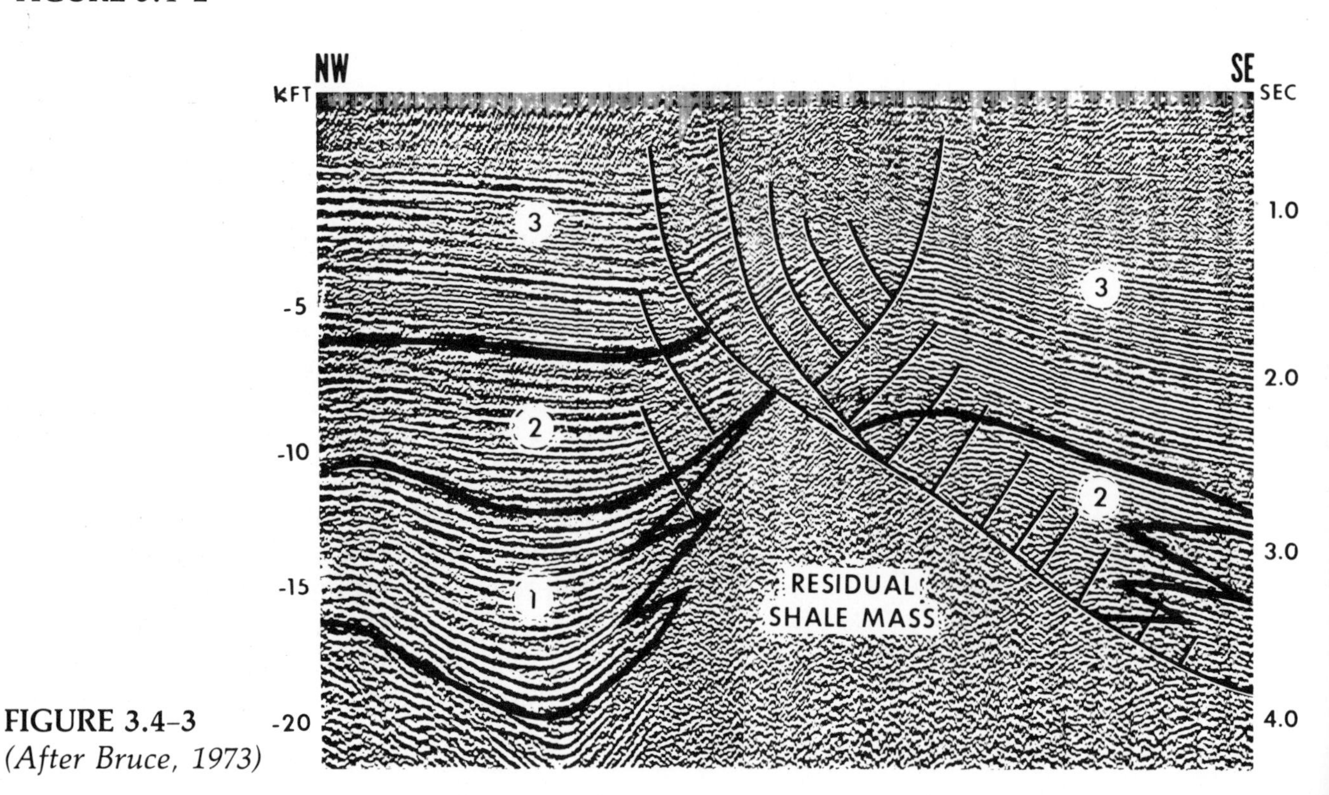

FIGURE 3.4-3
(After Bruce, 1973)

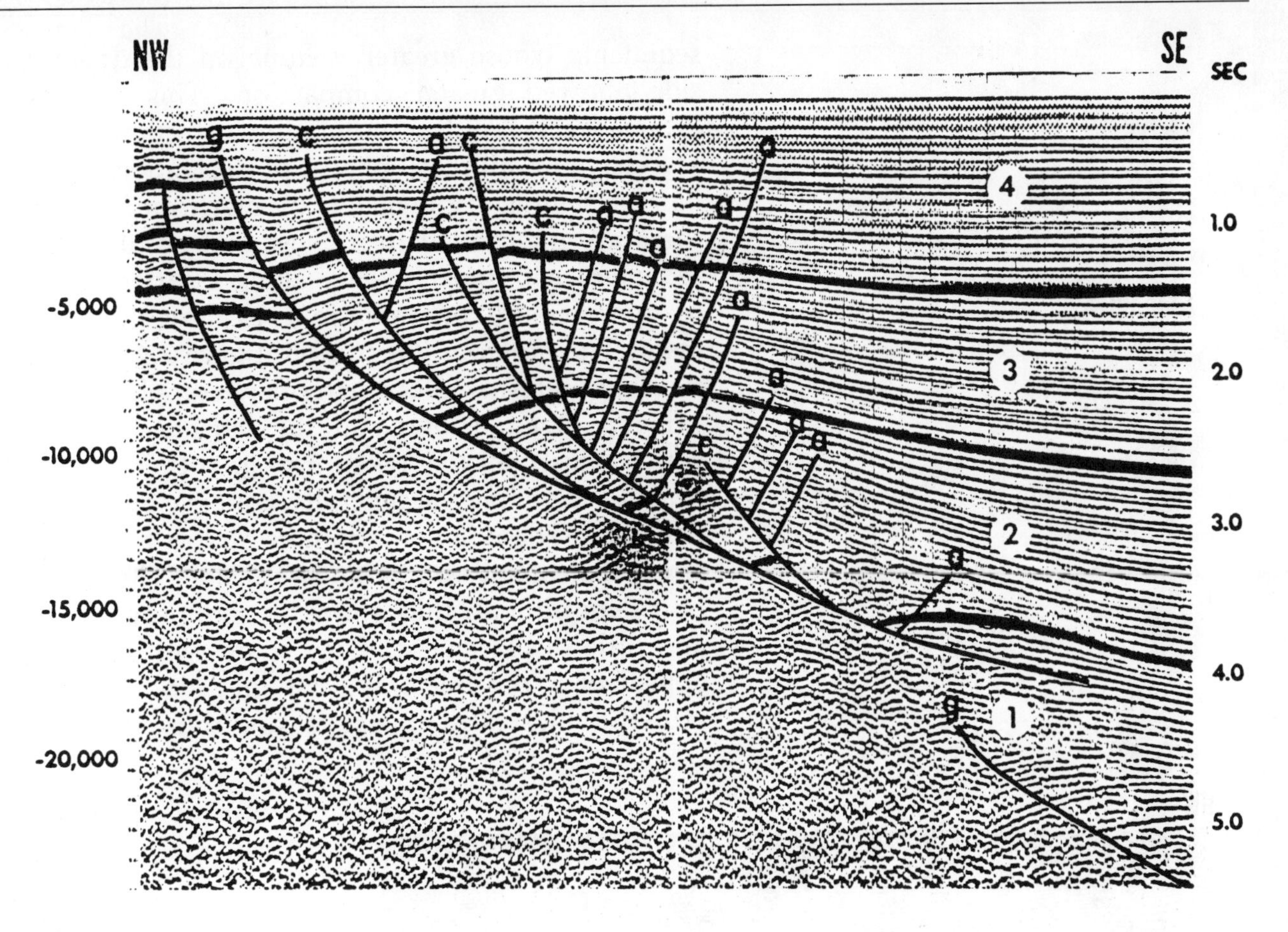

FIGURE 3.4–4

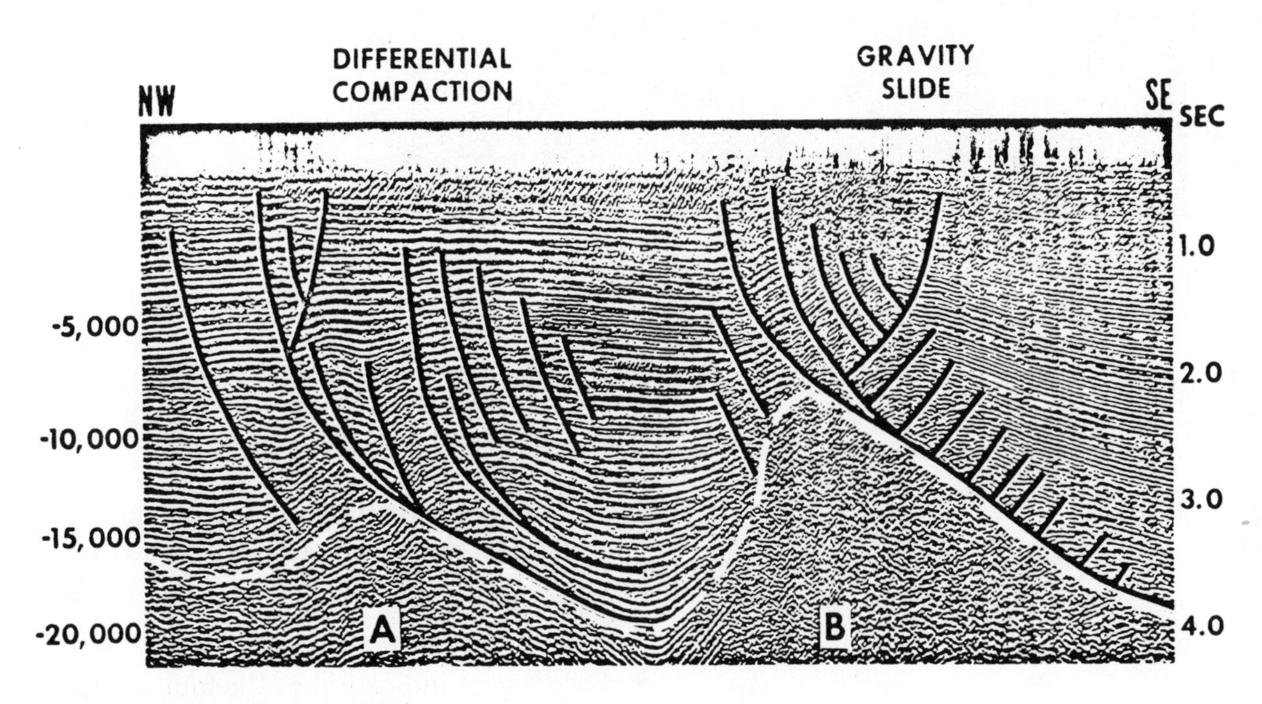

FIGURE 3.4–5 *(After Bruce, 1973)*

sediments whose greater volume on the downthrown side allowed greater compaction. This situation indicates very rapid accumulation of sediments during regressive or stillstand phases of deposition. The right-hand system is distinguished by many antithetic faults; these are interpreted to suggest that an abundance of sediments slid under gravity along a surface **lubricated** by overpressure, probably in a sea-floor-slope environment. The previous illustration (figure 3.4–4) combines both mechanisms in one extremely complex system.

Figure 3.4–6 shows, on the left-hand side, faults which must not be taken for growth faults merely because they have a generally concave-basinward appearance. These faults terminate almost at right angles to the shale mass—on its **landward** side. These faults are **post-depositional**; they show no significant thickening and little displacement. They share with all faults the appeal of a possible trap, but they do not share the growth fault's extra appeal of sand thickening.

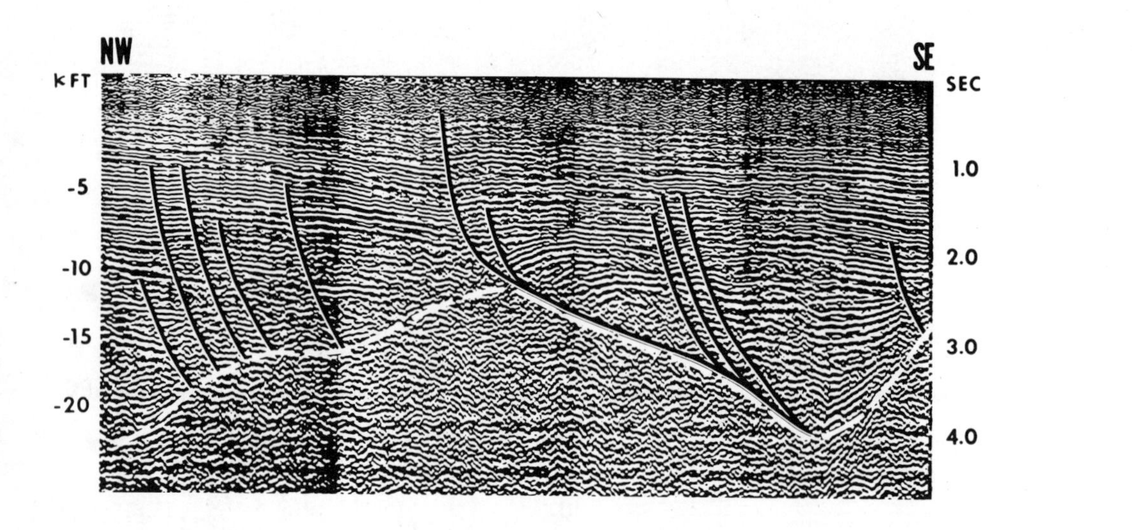

FIGURE 3.4–6

Figure 3.4–7, also from Bruce, indicates how the configuration of the faults can be interpreted to determine whether the depositional regime was regressive, balanced or transgressive.

There are a few features of the seismic technology which can affect the interpretation of growth faults very materially:

- Migration is essential in any complex fault situation.
- For clearest indication of the faults, and for effective migration, it is most desirable that the lines

FIGURE 3.4–7
(After Bruce, 1973)

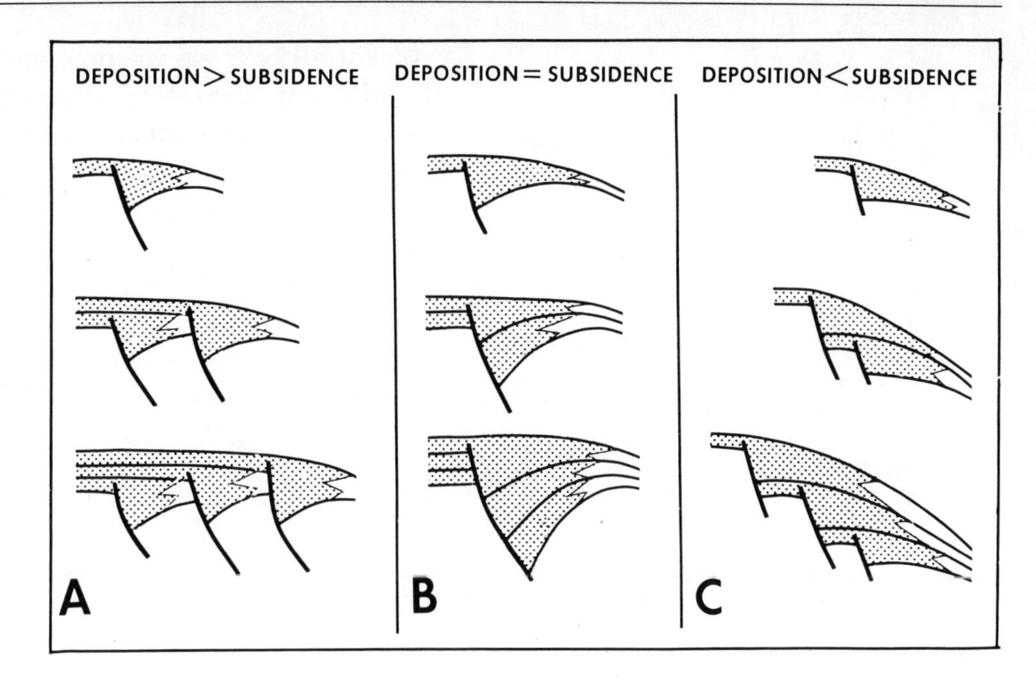

should be substantially at right angles to the faults. The migration of a line parallel to the fault trend is meaningless and confusing (and just plain wrong) unless it is part of a full three-dimensional migration.

- A complex system of faults requires more lines than would be sufficient to resolve simple structure.
- Time-variant processing operations contain an element of danger; the similarity of character of reflections which should correlate across the fault may be obscured merely because the reflections come at different times. Correlation across a growth fault is difficult enough without manufacturing additional problems.
- Similarly, there may be problems with the velocities. Not only may the velocities on the two sides of the fault be genuinely different (because of over-pressure, and different sand-shale ratio) but the velocities may be measured as different because of refraction and other transmission irregularities in the fault zone. Further, normal techniques for interpolating stacking velocities (vertically, and particularly horizontally) may not be apt across a fault zone.

- For all of these reasons, and others, the details of the reflection amplitudes associated with fault-trap reservoirs may not be very reliable.
- Where the seismic grid contains lines not perpendicular to the fault plane, fence-models made of seismic sections are invaluable (SITPA 608-619). A confused and uninterpretable zone on such a line normally becomes very reasonable and clearly explicable when the full three-dimensional picture is presented to the eye.

Although this section has concentrated on growth faults (because of their intrinsic connection with sand reservoirs), we should also add a few comments about post-depositional faulting and fracturing.

One comment concerns the contribution which post-depositional faults sometimes make to the identification of the lithology. On a properly-migrated section, such as figure 3.4–8, we can see a clear distinction in the tear behaviour at the fault; the shales (unless they are very old) flex into the fault, while the cemented sands and the carbonates snap. For equal thickness, the progression from snap to flex is the progression from quartzite through dolomite, sandstone and limestone to shale.

Another comment concerns our increasing appreciation of the importance of fracture porosity. An indifferent sand reservoir—a transgressive sheet sand, for example, or a delta-front sheet sand—may become a locally excellent reservoir where it is fractured; therefore part

FIGURE 3.4–8

of the exploration game is to find zones where the reservoir is fractured but where the seal (or a seal) remains intact.

Although earth tides and tectonic movement may provide the trigger, most natural fracturing is basically hydraulic; it occurs when the forces of pore pressure in a consolidated material become high with respect to the resolved horizontal forces acting in the rock framework. In seismic section, we are looking for evidences of **tension** or less-than-usual horizontal compression. Normal faults would be a positive indicator, as would be locally-reduced interval velocities or a clearly broken (not eroded) reflector; compressional folding would be a clear negative indicator. Where extensional folding or flexing occurs (for example, over a rejuvenated uplift), the maximum fracturing is found in the thickest beds and at the maximum **rate-of-change** of dip; it is the curvature of the beds, and not the dip itself, which is important.

Many or most extensive fractures are near-vertical; they almost always show a preferred direction or *grain*, and two (or more) intersecting grains are common. The directional grain(s) can often be identified by studying the lineaments visible (to a trained eye, or to a squinting doddlebugger) on LANDSAT images, U-2 photographs, or SLAR displays. Indeed, some exploration departments now predispose their drilling of sheet sands according to the presence or density of fracture lineaments made visible by these means at the surface.

3.5 Fluvial sand bodies

We turn now to the types of sandstone reservoir which are in their nature **local**—channels, fans, beaches and the like. An important criterion for recognition of most of these targets is their **shape**.

Let us start on land, and wend our way down to the sea.

3.5.1 Alluvial fans

Figure 3.5.1–1a illustrates the environment of a typical alluvial fan; sketches (b) and (c) are sections through AA′ and BB′.

Alluvial fans are important as reservoirs both in the form of single fans (the Quiriquire field in Venezuela)

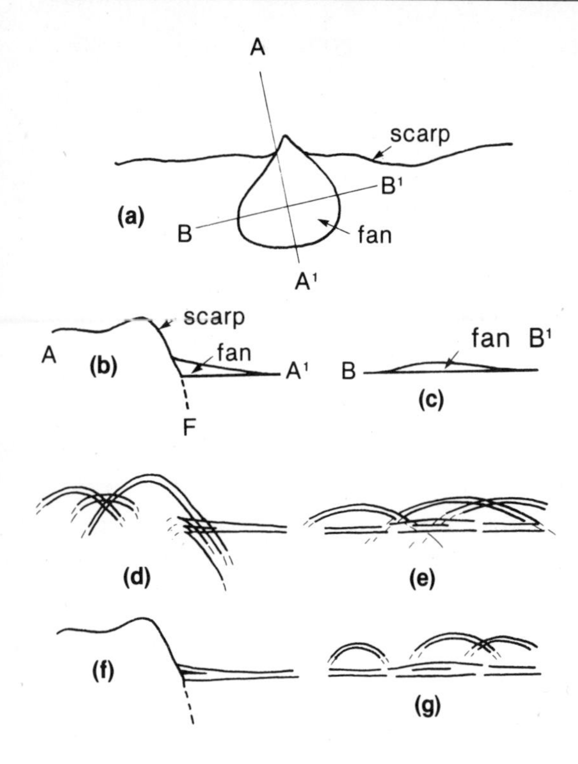

FIGURE 3.5.1-1

and in the form of stacks of fans coalescing together to form a thick wedge of sediment against a mountain front. In the latter case we have no help from the fan **shape**, but we can hope to recognize the wedge of non-marine sediments— variable amplitude, poor continuity—in its setting against the mountain front.

The interior of an alluvial fan is likely to contain everything from boulders to clay, with many old stream courses within it. Locally there may be systematic transitions of grain size, sorting and porosity, but at seismic scale the fan is likely to be just a jumble. Our hopes of seeing a single fan then depend on its areal size, its thickness and its differentiation from the surrounding materials. If we are successful, we still have the caution that the fan is likely to contain many small and possibly unconnected reservoirs, individually indistinguishable to the seismic method.

But could we see a single fan?

If it is very porous, gas-saturated, and capped by shale, the answer is probably yes—it may give a good bright-spot, and we shall be able to estimate its thickness and map its areal form.

If it is almost totally cemented, and liquid-saturated, the answer is again yes—but it then has little exploration appeal.

If it it porous and liquid-saturated, the conclusions of section 2.2 warn us that we may have little or no reflection from the top of the fan. If there is a weak reflection, there is a chance we could map its areal form, seeking confirmation by monitoring the **change** of reflection strength from its base. If there is no reflection, we cannot detect the fan directly.

The question then arises: can we, by studying the environment, identify situations where fans are likely to be—whether or not we can see them directly?

The first key to such identification, obviously, is a prominent scarp—probably a fault scarp—above the sea level of the time. Then we require evidence of rising sea level, so that fine-grained marine sediments may provide a seal. So our initial concern is to recognize, in seismic section, a faulted erosional surface and a rising sea level; these, fortunately, are often fairly straightforward (SS 63-82).

Having found the right environment, but still finding no direct evidence of a fan, we might screw up our courage and look at the diffractions.

Diffractions? Did someone say **diffractions**?

On a seismic line AA′, we expect to see diffractions from the mountains and valleys of the scarp. On an unmigrated section these diffractions may be strong, and the scarp face itself may not be clear. Migration should collapse the diffractions and (if the migration program can handle the dips) restore the steep scarp.

Let us go back for a moment to the situation of the gas-saturated bright-spot fan. Then diffractions are also likely from the scarp edge of the fan. In this case the section might appear as sketch (d) of figure 3.5.1-1, and the migrated section, very satisfactorily, as sketch (f).

The corresponding sketches for seismic line BB′ are given in sketches (e) and (g); these sketches contain no surprises at the level of the fan, but the eye is caught by the diffractions **above** the fan. These, of course, are the diffractions from the scarp; they are observed sideways, which is why they are not collapsed by two-dimensional migration. If the scarp edge is uniform, the diffractions appear as a continuous "reflection" whose time depends on the distance of the line from the scarp and whose apparent dip depends on the angle between the line and the scarp. If the scarp is irregular (as it must be, to have generated a fan) the diffractions represent that irregularity. Here then, is the key: there is just a chance that—even in the absence of reflections directly from the fan—we could interpret these diffractions to give us the position of a **valley** leading to where a fan must be.

This sort of approach is a long shot, of course, and probably appropriate only after one such fan has been found. The diffractions are the key—not much, but all we have. We have to admit that it would need a lot of courage to propose a location on a seismic section which showed a lot of nothing and a couple of diffractions. Well, no, perhaps not; diffractions have been drilled before now—on top. But to drill **between** the diffractions...?

3.5.2 Braided-stream deposition

Downstream of the alluvial fan, the river may take the form of a braided stream, flowing across a plain of moderate slope. A ridge of mountains, generating a number of fans and braided streams, may thus give rise to an extensive area of braided-stream deposition—characterized by variable but locally excellent porosity

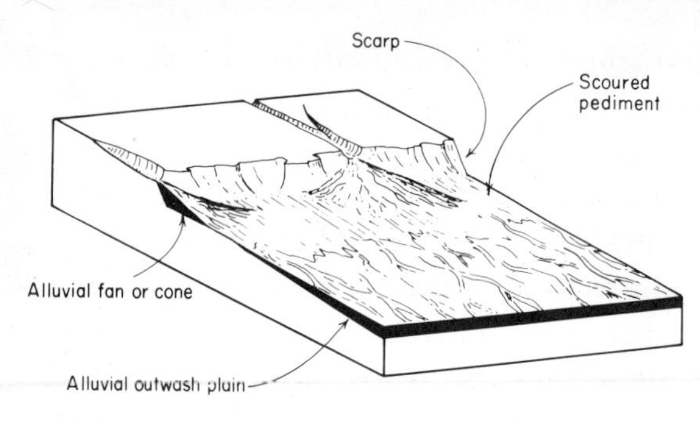

FIGURE 3.5.2–1

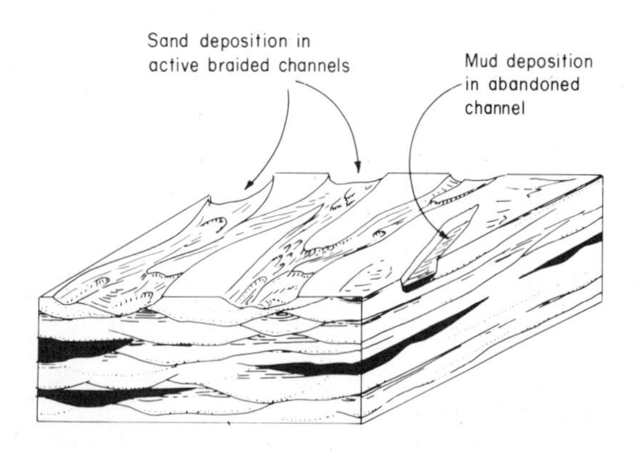

FIGURE 3.5.2–2 *(After Selley, 1976)*

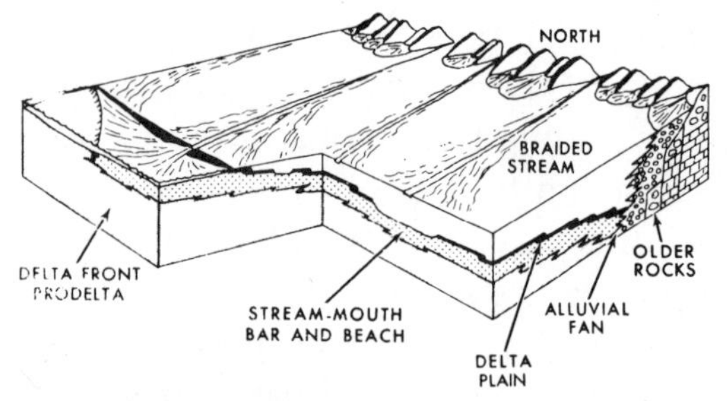

FIGURE 3.5.2–3
(From Eckelmann et al., 1975)

(figures 3.5.2–1 and –2). A modern example is the Canterbury Plain in New Zealand; a dramatic ancient example is the Sadlerochit Formation at Prudhoe Bay, which includes as much as 150 m (500 ft) of prolific braided-stream reservoir (figure 3.5.2–3; Eckelmann, Dewitt and Fisher, 1975).

There is not much in the nature of such a reservoir which would make it look particularly interesting (or even obvious) on a seismic section. 150 m of braided-stream deposits would look like 100 ms of variable-amplitude-and-poor-continuity—the standard response of the seismic method to non-marine sediments (SS 169). The only distinction we can expect is that the variability is likely to be less marked than from the equivalent thickness of alluvial-fan deposits. Again, it is in the recognition of the **total** environment that we can hope to identify positively the prospective nature of such a unit; in figure 3.5.2–4 we would be looking for a ridge of mountains (a), a possible alluvial fan (b) or a stack of such fans, a non-marine zone (c) which we hope is a stack of braided-stream deposits, a marginal-marine situation (d) (perhaps including a delta), the signature of a prograding slope (e) and its proof of sea level at the time, and evidences of subsequent transgression (f) by which the whole system was preserved.

3.5.3 Stream-cut channels

Before reaching the alluvial fan, or in later passage through an uplifting area, or as its gradient is increased by falling sea level, the river is likely to cut valleys, or even canyons. Here we are concerned with these channels, cut by streams of steep-to-moderate slope. In cross-section, small channels may be triangular; however, the large mature channels are likely to be flat-

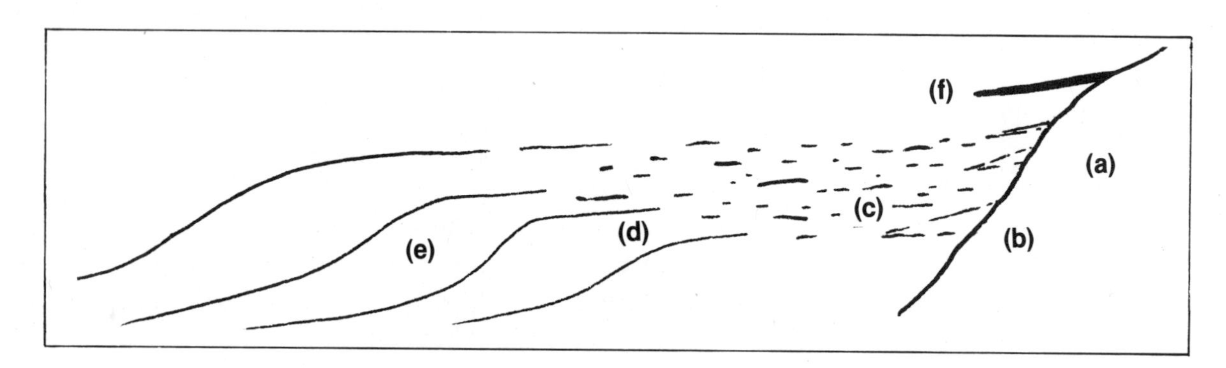

FIGURE 3.5.2–4

bottomed, sometimes with fairly steep sides. In plan, we expect the channels to have the obvious long-and-narrow form; in a watershed area we also expect a characteristic drainage pattern—dendritic, trellis or rectangular.

Our first interest arises when such channels become filled, partly or completely, with reservoir material. This can happen by reason of a change in sediment load, or in river volume or grade, or by back-up from the sea, or as a late stage in erosional maturity of the landform. Then the channel itself becomes our potential reservoir, and the object is to drill into a favourable part of it.

Our second interest arises when the channel, substantially cleared of coarse-grained material, becomes filled with clay during a subsequent transgression; then the channel is the seal, a material truncated by the channel is the reservoir, and a bend in the channel (or the channel plus tributaries) is the trap. The second situation is therefore a special case of an unconformity trap.

Some features of these two situations are common—both advantageous and disadvantageous. Let us deal with these common features first.

- As always, we must ask whether the acoustic contrasts are sufficient to yield detectable reflections. We can hope for clear reflections if the channel is cut in limestone, or if we have a very porous gas-saturated reservoir, or if we have a hard tight sand somewhere; otherwise we must make our usual resignation to fairly weak reflections.
- Again, as always, we must ask about the thickness of the channel. There are famous examples of channels whose depth is a hundred metres or more, but many more good ones remain for which seismic resolution is marginal or inadequate.
- Then we must ask about the width of the channel (as we did in section 2.6). In many practical cases the width of the channel is less than the diameter of the reflection zone; then the reflection strength we see does not properly indicate the acoustic contrast present.
- Shall we see diffractions from the sides? If the channel is like the Grand Canyon—cliff-sided—then, yes, we shall; if the channel sides slope gradually, then no.

- If the channel is wide enough, and reasonably flat-bottomed, we have a fair chance of seeing a reflection from the base. Of course this depends on the actual materials, but at least we do know that the interface is abrupt.
- The most powerful direct indicator of a channel, on a seismic section, is a zone in which the continuity of parallel reflections—even just the shale grain—is interrupted. The walls of the channel may not show as reflections, nor even the floor of the channel, but the limits of the channel are clearly shown by the break in the grain. Occasionally we may see other reflection alignments within the channel (in some young examples, with glacial fill, the four ice ages are visible), but these alignments are obviously foreign to their surroundings. Figure 3.5.3-1 (from SITPA 576) illustrates at large scale the features we are seeking: the total break in the continuity of intersected reflections, the obviously foreign nature of the sediments and reflections within the channel, the reflection from the base, reflections from the sloping sides and diffractions from the cliff sides. (Incidentally, the channel fill in this case is **high**-velocity glacial detritus, accounting for the entirely spurious uplift observed below the channel.)

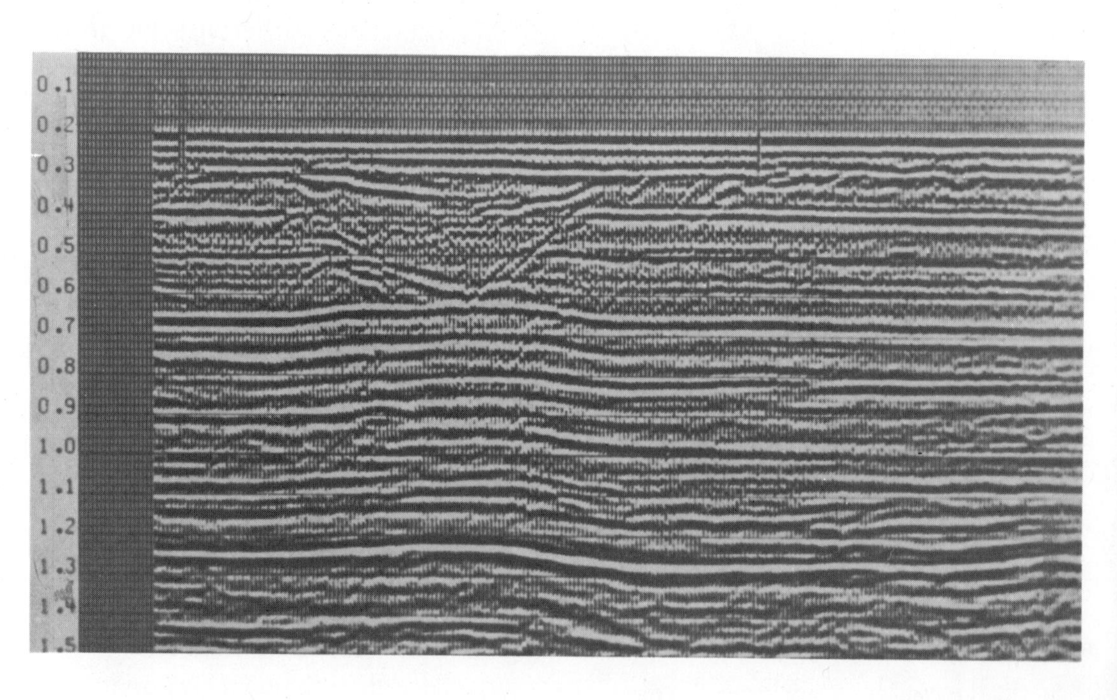

FIGURE 3.5.3-1

- If the reflection grain interrupted by the channel is substantially parallel to that above the channel, the channel was cut in horizontal beds; in general, it can be expected to be part of a **dendritic** drainage system (Busch, 1974, p. 96). The importance of recognizing the type of drainage system is that it gives guidance on the angle at which the tributary channels may be expected to join the main channel; in a dendritic system the tributaries join at a fairly acute angle.
- To make use of this, we need to know the direction of flow in the main channel. If we have regional control this is no problem; otherwise we search for the direction of thickening of the sequence into which the channel is cut. If the bottom of the channel is visible seismically, we look for thickening of the channel itself, or map the interval between the channel bottom and a suitable marker **above** the channel. We always insist, of course—as the subsurface geologists do—that the signature of a channel demands down-cutting relative to a datum above.
- An alternative approach, applicable to all types of drainage patterns, is applicable whenever one of the reflectors breached by the channel is particularly clear; this occurs, for example, in many cases where the channel breaches a limestone bed. Then it is merely necessary to map the terminations of the limestone reflection.

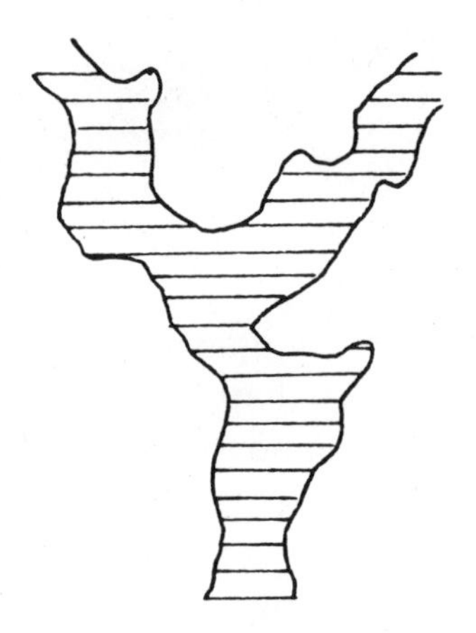

Now let us return to the reservoir possibilities of these channels.

The first possibility, we said, was that the channel had become filled with sediments including reservoir material.

At large scale, this can occur when a major river channel becomes inundated during a rise of relative sea level, with the river continuing to supply a considerable volume of mixed sediments. As the river grade reduces, this suggests that the transitions of the river—from fast-flowing stream to braided stream to meandering stream—move back up the valley, and that the possible environments for accumulation of coarse-grained sediments move likewise. The upper section of figure 3.5.3-2 shows the upper reaches of a very large channel off the coast of Yugoslavia, and the lower section shows the middle reaches. In the middle reaches we see some

FIGURE 3.5.3-4

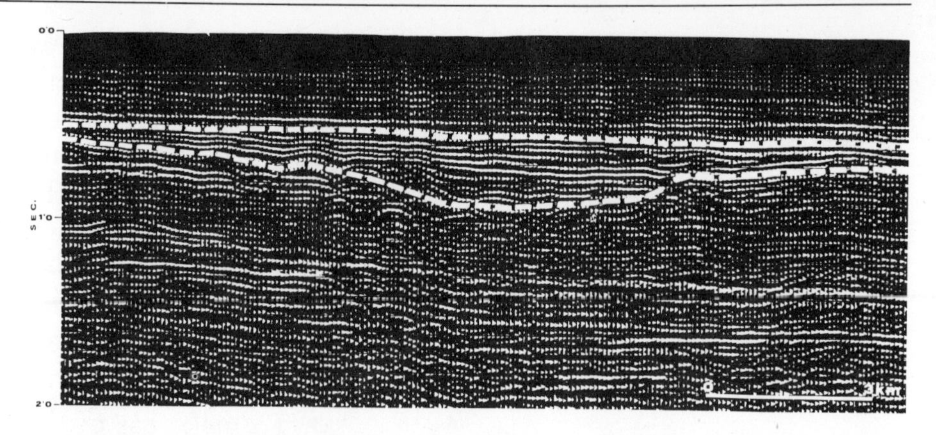

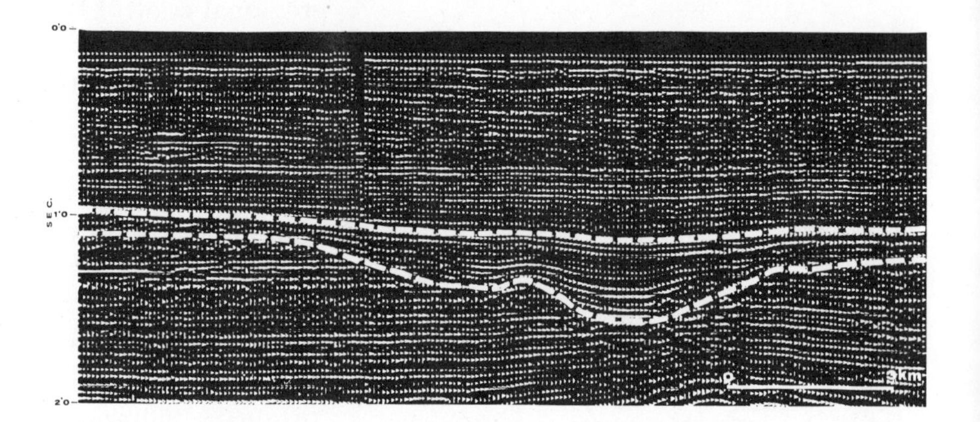

evidence for sands (believed to be gas-saturated) near the base of the channel; the internal appearance and obvious compaction of the material above these sands suggest a shale cover. In the upper reaches we see contrasts of acoustic impedance virtually filling the channel. If the conditions for generation and sealing of petroleum were favourable, there are good hopes of discoveries in meander-type reservoirs near the base of the lower channel, and in braided-stream type reservoirs throughout the thickness of the upper channel. Viewed as a whole, this type of backed-up channel appears to be a significant exploration play.

At a smaller scale, we must consider the situation where a stream-cut channel becomes filled with local erosional detritus (perhaps after a major climatic change, and the drying-up or diversion of the river). As the topography matures, the channel fills—normally with an unsorted mix of local detrital fragments. Ordinarily this may not be very interesting, but a reservoir situation may develop (particularly if there should be a measure of wind-blown transport).

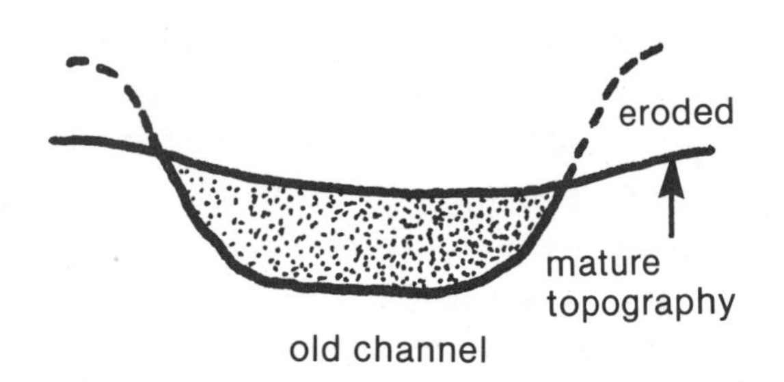

We would expect to find the seismic signature of this type of prospect on any mature erosional surface, and indeed most such surfaces on seismic sections show features which can be interpreted as filled channels. We ordinarily ignore these evidences; perhaps we should be looking at them more closely.

How could we do that? First, by satisfying ourselves of the maturity of the erosional surface, of the position of relative sea level at the time, and of a subsequent marine transgression to provide a seal. Then, by proving that this was an erosional channel filled with detritus by noting the absence of shale compaction drape over the feature the ("the post-compaction cut-and-fill" of Busch; 1974, p. 74). Then, by plotting the configuration of the channel in plan (an undertaking, incidentally, which is likely to be expensive in closely-spaced seismic lines). Then, by studying the subsequent regional tilting of the area, and estimating the manner in which bends and high points of the channel have become traps. And finally, by looking for changes in reflection strength which might suggest gas in those updip traps.

Of course, as we look at smaller and smaller features we shall become frustrated by inadequate channel thickness and inadequate channel widths. The seismic response will become so subtle that we could attempt to interpret it only where we have borehole control. Nevertheless, it does seem to be true that we could investigate the larger channels more often than we do.

The erosional channels we have been discussing are distinguished from deltaic distributary channels (for which the sediment source is the river itself) by the absence of shale drape over the fill; we shall consider the deltaic situation later. However, there are many intermediate types of channel fill, and we might identify in particular the Red Fork sandstone reported by Lyons and Dobrin (1972). In this case the river had cut a channel into a shale which was still capable of significant compaction, and had probably contributed itself to the transport of the filling sand. Because of the greater compaction of the youngish channel-cut shale (relative to the channel-fill sand), the subsequent capping shale and limestone became draped over the sand. Thus the position of the channel was found by mapping the time interval between good reflections above and below the channel (figure 3.5.3–3). The work was done many years ago; the old records shown in the figure would en-

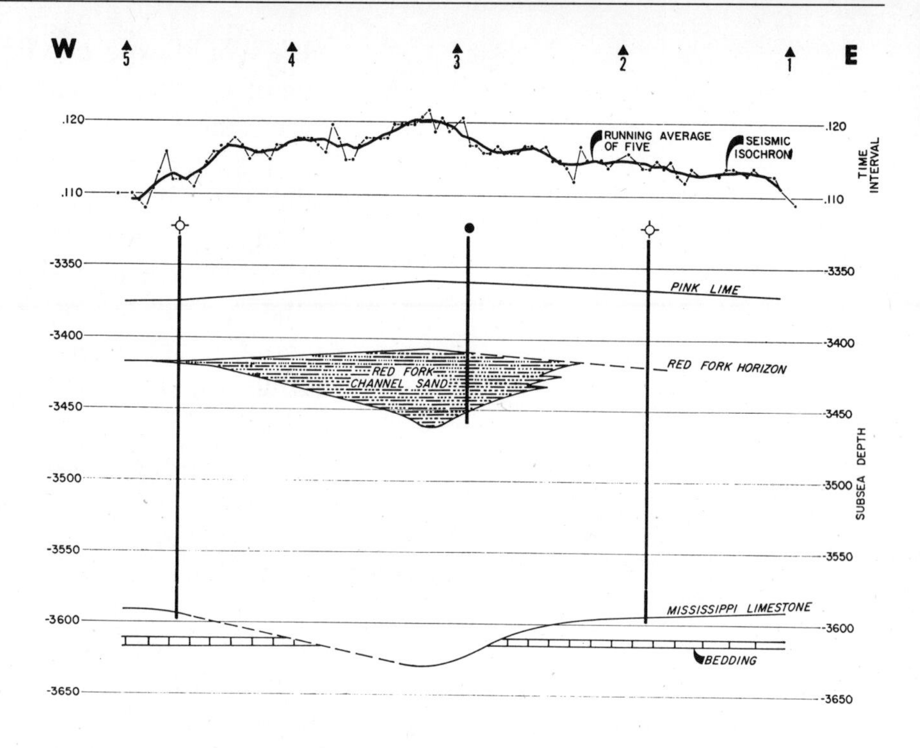

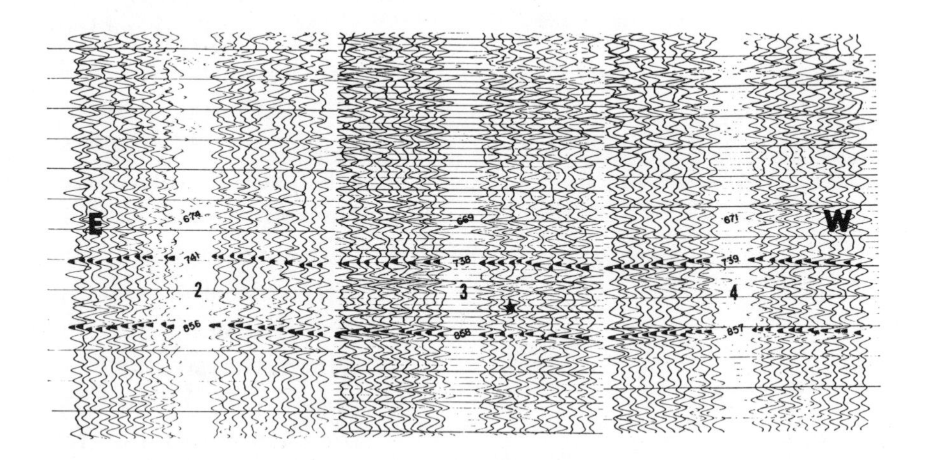

FIGURE 3.5.3–3
(After Lyons and Dobrin, 1972)

courage us to hope that with today's technology the channel could be clearly seen and mapped as a reflection feature, once its signature became known. It still remains true, of course, that the time-interval mapping is sound practice,* but it can find only those channels characterized as "pre-compaction cut-and-fill" (Busch,

* It is most important that the isopach map of an interval containing a channel should not be constructed by machine, unless the machine has been programmed to respect the particular (and non-structural) constraints associated with a channel; a normal machine-contouring program must not be used.

FIGURE 3.5.3–4
(After Lyons and Dobrin, 1972)

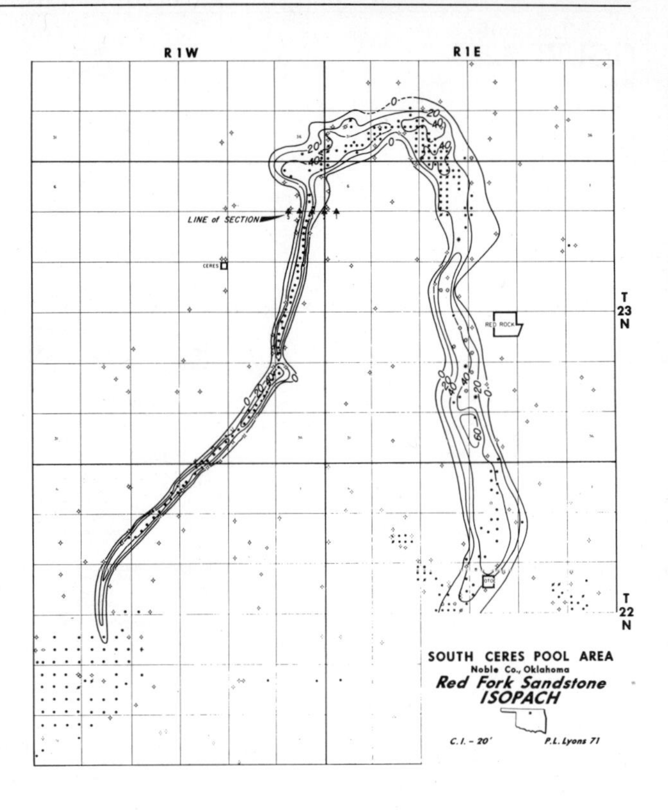

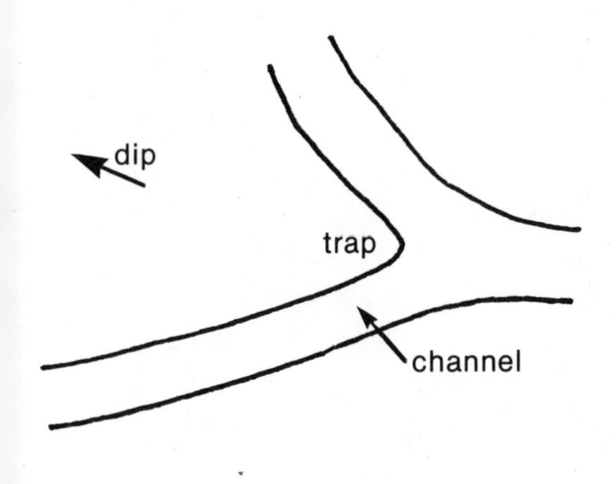

1974, p. 74). In the final analysis, it is the configuration of a reflection measurement in **plan** view (figure 3.5.3–4) which establishes the prospect as one or more river channels.

Now let us consider the second type of channel reservoir—where the channel itself is full of **clay**, and the clay provides an updip seal and trap. The situation we seek, in seismic section, is that of figure 3.5.3–5. Here the channel is identified primarily by the interruption of the breached reflectors, and by their replacement with a rather uniform material having the appearance of shale. Left of the channel the breached reflections show leftward dip, which means that we have the possibility of an updip trap if the breached sequence contains a reservoir sand.

Once this situation is observed, then, we search for the trap in plan. This we do by mapping the updip limit of the breached reflections against the channel, and by attempting to find some combination of tributaries and bends which will complete the trap.

Such reservoirs are the standard exploration objective in the Upper Minnelusa sand of the Powder River Basin. With modern seismic techniques, the clay-filled channel

FIGURE 3.5.3–5
(After Farr, 1976)
(Courtesy Western)

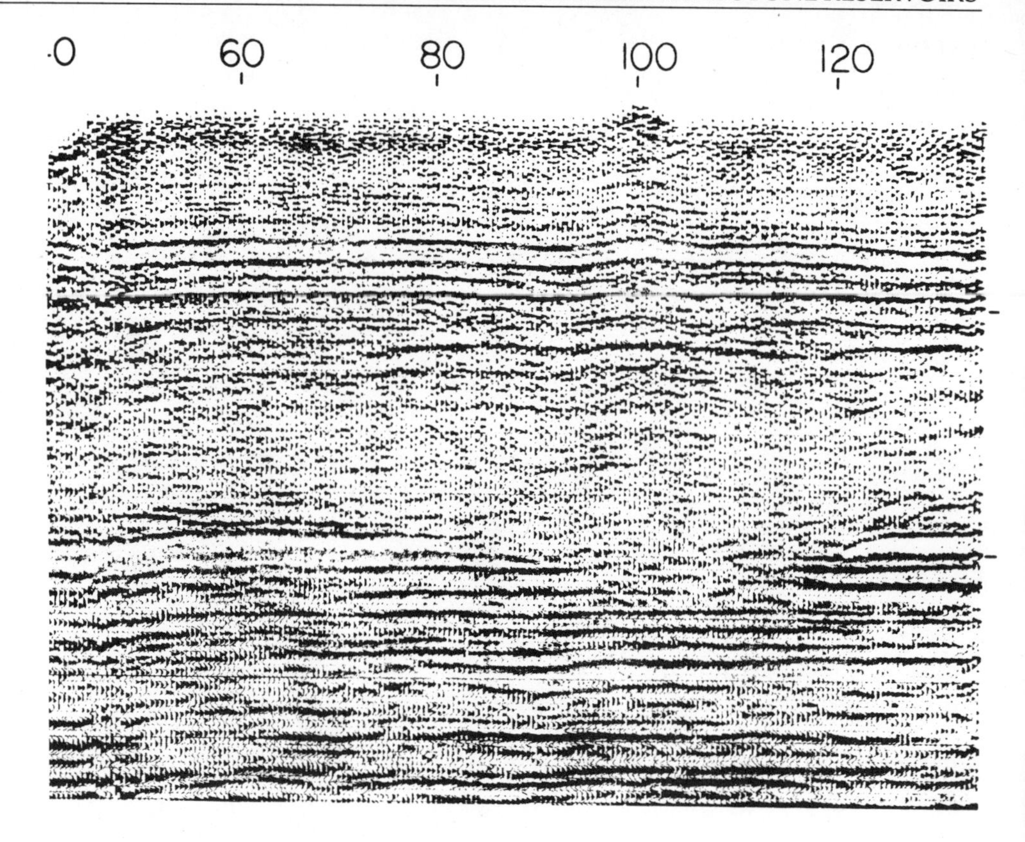

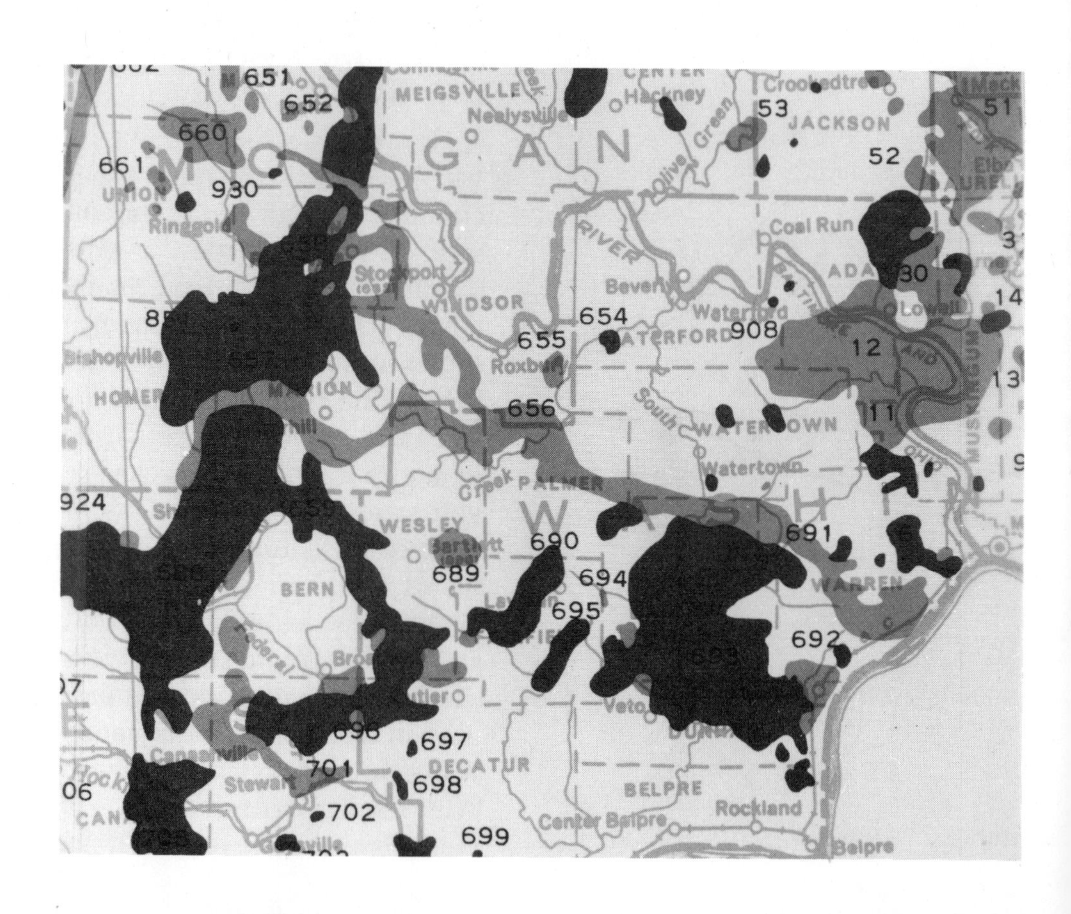

FIGURE 3.5.3–6

(typically 1 km wide and 30 m or 100 ft deep) is sufficiently clear to allow accurate mapping. The same mechanism at larger scale accounts for the important Ubit field offshore Nigeria (Weber and Daukoru, 1975).

We scarely need to note one of the side-attractions of channel-fill prospects: if there is a channel, there must be the rest of it. Figure 3.5.3-6 shows a small portion of the oil and gas map of Ohio; we have only to look at the plan view of the WNW-ESE field to see its origin as a channel, and to guess where extensions of the field might be sought.

3.5.4 Meandering streams

Now let us follow our river to its maturity—to the broad flat plain across which it meanders lazily, before its final reckoning with the sea.

What are the characteristics of this situation, from a seismic point of view?

Individual meanders generate point-bar sands on their inner sides (figure 3.5.4-1). As the process continues, a meander belt is generated; this belt is ordinarily much wider than individual meander loops (figure 3.5.4-2). In the mature state, a large part of the meander belt may be occupied by point-bar sandstones, and the system may build upon itself so that both horizontal and vertical multiplicity of sands becomes possible. The inner edge of each bar is transitional, while the outer edge is likely to be an abrupt contact with a mud-filled channel. The channel itself is of no interest, except as the outside seal of the point-bar; the target is the arcuate sand within each meander of the river (figure 3.5.4-3). The lower contact of the point-bar is abrupt; the upper contact (because of a fining-upwards transition within the sand) is likely to be less strongly marked.

Seismically speaking, individual point-bars are thin (perhaps 10-20 m; 30-70 ft). Areally they are not large either (seldom reaching the possible 5 km or 3 miles across). However, a vertical **stack** of point-bars in a meander belt can be significant in both thickness and width.

So do we have a seismic signature?

In terms of the environment, we know that the meander belt must be on low-lying flat ground (probably in a coastal plain or a very broad valley), and that it is likely to be approximately perpendicular to the gross shoreline of the time. Further, the discovery of a delta on a seismic section clearly points the finger to a

FIGURE 3.5.4–1
(After Le Blanc, 1972)

flat coastal plain on which a meandering river is likely.

In terms of seismic response, we have to accept that the reservoir zone may not be very obvious if the sand has good porosity and is liquid-saturated. We can expect some fragmentary reflections developed at the point-bar contacts, but these reflections are seen against a background of adjacent and subjacent non-marine deposits which also return reflections of variable

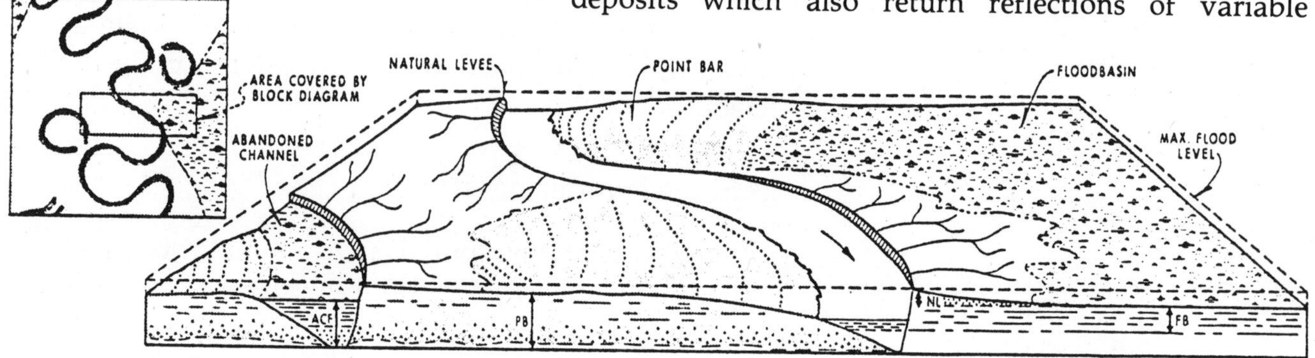

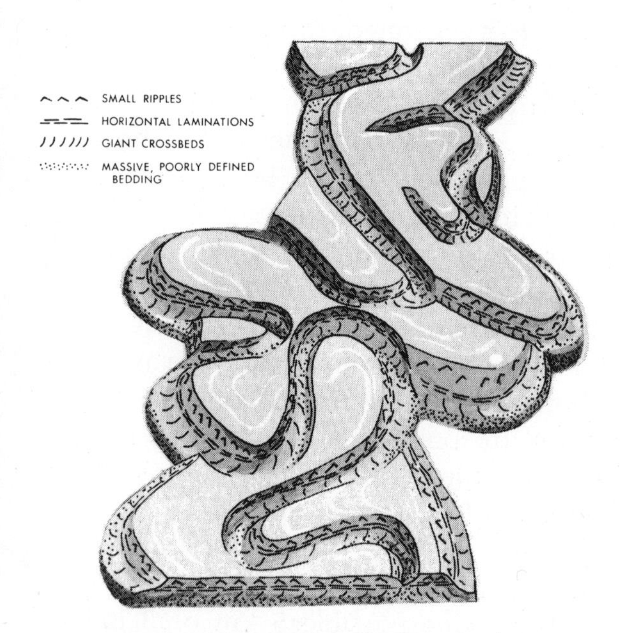

Drawing of model of hypothetical meander belt, showing how a complex point-bar reservoir sandstone would appear with clay plugs and surrounding material stripped away. Distribution of bedding structures is generalized. Zonation tends to follow this arrangement for a complete sequence. One or more of upper zones can be eroded away during later stages of meander development and a new sequence (or sequences) deposited on top. Plaster model prepared by John R. Warne (personal commun., 1967).

FIGURE 3.5.4–2 *(After Busch, 1974)*

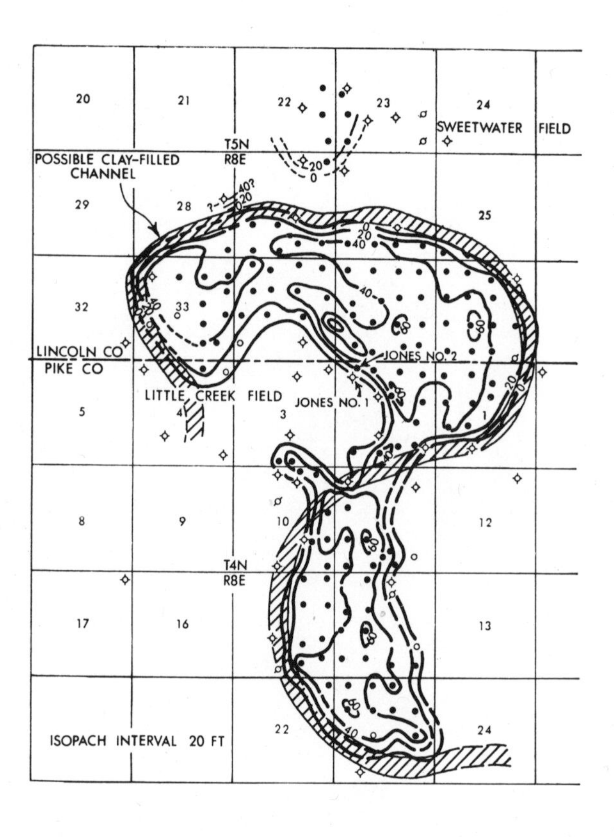

FIGURE 3.5.4–3
(From Busch, 1974; after Eisenstatt, 1960)

amplitude and fragmented continuity (SITPA 421). We cannot expect the point-bars to generate a very characteristic signature against this background.

However, if the sand porosity is low enough to give us usable shale-sand reflections, and the sand is not too thin, we begin to have a chance. The appearance of a point-bar on a single line would be uninterpretable, of course; it is the **areal mapping of the reflection amplitudes** which gives us the key.

A delightful example is that of figures 3.5.4–4 and –5 (Lindsey, Dedman and Ausburn, 1978). The first figure shows the relevant part of the seismic section on one line across a meander belt; the level of interest is between the two continuous reflections—where a reflection can be seen to come and go. It is the strength of this reflection which is to be mapped areally.

On the left of figure 3.5.4–5 is a subsurface reconstruction of a point-bar oil field, showing the location of the seismic line illustrated above. On the right is an interpretation of the seismic amplitude data. The general duplication is most impressive; where there are differences they are interpreted as being due in part (as we would expect) to the finite dimensions of the reflection zone. As we discussed in section 3.1, actual quantification of the thickening and thinning of the sand (and so the first estimate of reserves) can come only from amplitude calibration or from borehole control.

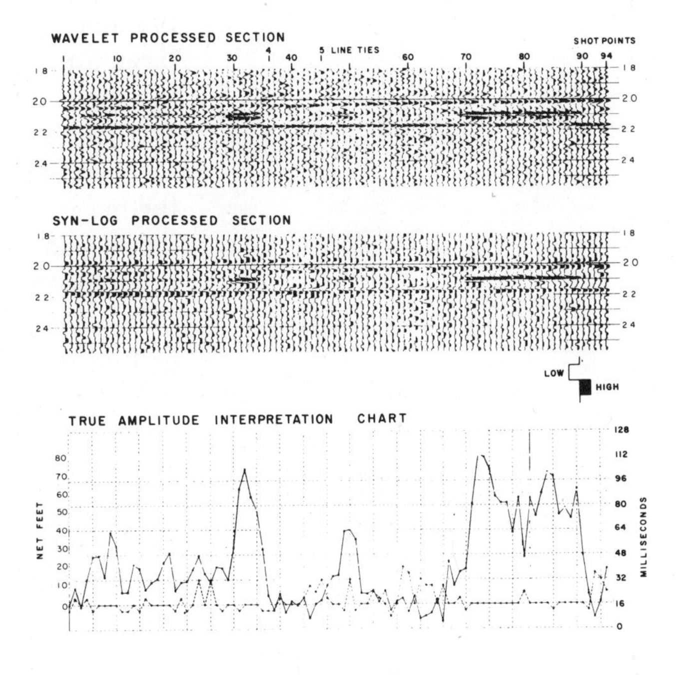

FIGURE 3.5.4–4

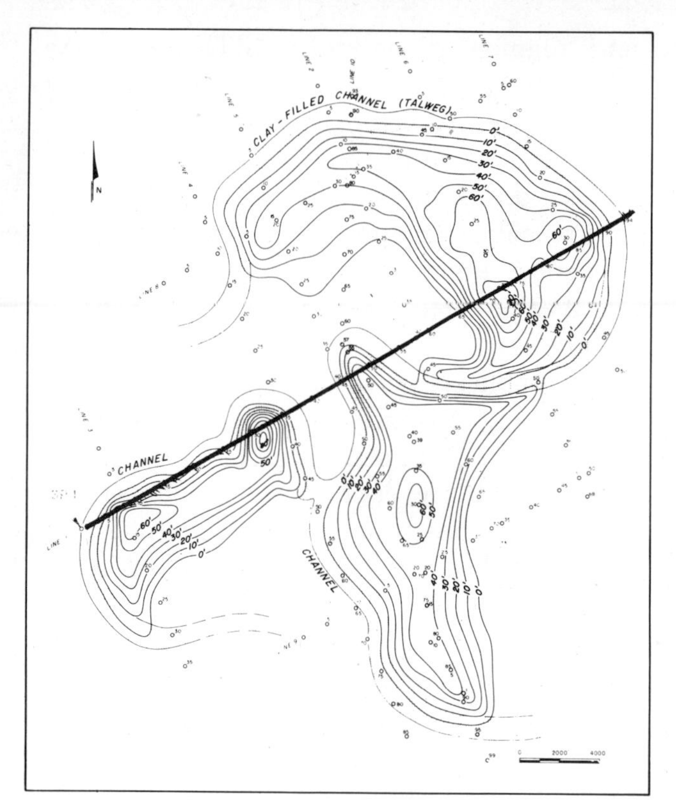

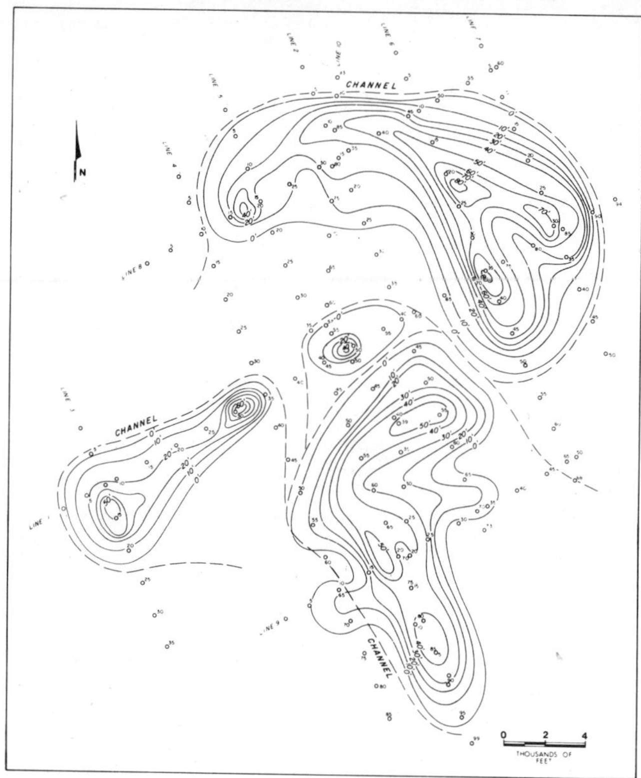

FIGURE 3.5.4–5 *(After Lindsey et al., 1978)*

In the case where the saturant is gas, and the point-bar sand is of good porosity, we can expect to see very large amplitudes. However, complications arise in the interpretation of the mapped amplitudes, because if the system has become tilted the strong amplitudes may no longer identify with a point-bar pattern. This means a further stage of interpretation, reconciling the observed amplitudes with the direction and degree of tilt and the orientation of the suspected meander belt.

Whereas we have said that the final channel delimiting a point-bar is not of interest itself as a reservoir, the same is not true of abandoned channels forming ox-bow lakes. There is a fall in river velocity associated with the development of the short-circuit across a meander, and this often leaves sand in the near part of an ox-bow lake. Such a target, however, is likely to be too small for us; we would be better advised to concentrate on the point-bars.

3.6 Deltas

A comprehensive general reference is Coleman (1976).

We have already considered, in section 3.2.3, the recognition of a delta by its characteristic oblique-progradational appearance, and the identification of its delta-front sheet sand. We recall that it is the oblique feature of the signature which raises our hope of a high-energy balanced situation (SS 123-128). Often, we see the oblique signature **within** the sigmoid (lower-energy) signature (figure 3.6–1); this may occur on different seismic lines at different levels, telling us (within an uncertainty associated with subsidence and changes of sea level) something about the movement of the delta up and down the coast. All these evidences are clearest, of course, when the delta front is prograding into deep water, but even in shallow water we can often recognize lobes of outbuilding by a combination of the correct environment and a **hummocky** reflection appearance (figure 3.6-2).

0.60 SEC.

FIGURE 3.6–1

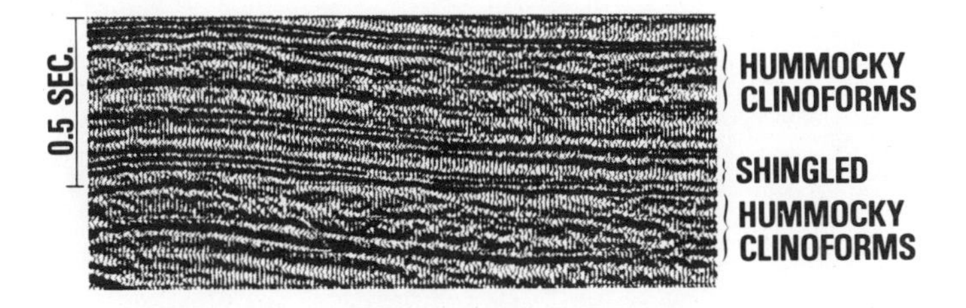

FIGURE 3.6–2 *(After Mitchum et al., SS)*

The identification of a location where we can expect to find delta-front sheet sands is almost routine, therefore, and sometimes even easy. But these sheet sands are not clean, in their pristine state, and consequently not the best of exploration objectives. Far more desirable, in a reservoir sense, are several specific types of sand bodies which can form as integral parts of the delta system—particularly distributary-channel fills, bar-finger sands, and delta-margin islands.

Because there are several distinct types of delta—depending on the balance between the river, the tide, the depth of the water, the longshore currents and gravity—a particular delta may not generate all of these sand bodies. However, these distinctions between types of delta are unlikely to be evident in seismic section—with two exceptions.

The first is that with a good grid of lines we should be able to determine whether the delta is of pronounced lobate form, or whether there is comparatively little outward bulge of the shoreline in the delta region. In the first case we are looking for our good sands within the delta and on its seaward edge; in the latter case we are expecting that longshore currents have moved the best reservoirs to bars and beaches along the coast.

The second distinction which may be evident in seismic section is the depth of water into which the delta was discharging. If the water was shallow our most likely targets within the delta are the distributary channels; if it was deep they are the bar fingers.

But do we have any chance of seeing such individual reservoirs? Let us review their characteristics, and decide.

(1) Abandoned distributary channels may become filled with sand, with the coarsest sediment tending to be at the upstream end. They may form an extensive braided

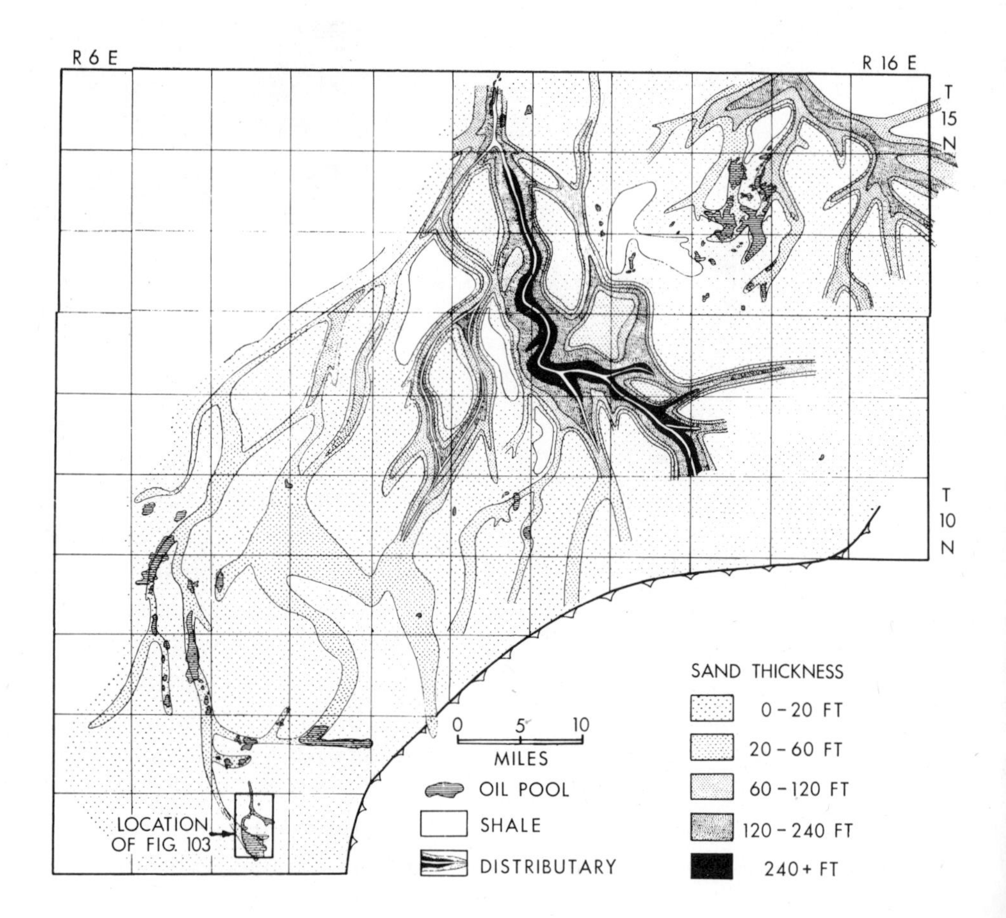

FIGURE 3.6–3
Isopach map of Booch sandstone, greater Seminole district, eastern Oklahoma.
(From Busch, 1974)

pattern, tens or hundreds of kilometers across, but individual channels may not be very wide. Our immediate reaction to this is that the seismic method would be powerless to detect individual channels, though the braided pattern as a whole might show as a local amplitude-and-character anomaly—particularly if the sands contain gas. However, there are several cases in the literature where we could hope to see individual channels, and to build up from them a picture of the delta as a whole. The best example is the Booch* delta of eastern Oklahoma, where the major channel is some 56 km (35 miles) long, 5-8 km (3-5 miles) wide, and at least 73 m (240 ft) thick. Figure 3.6–3 is taken from Busch (1974, figure 102). Several other ancient examples (including the central Pennine trough in England, the Bell Creek delta on the Wyoming-Montana line, and the famous shoestring sands of the Kansas-Oklahoma line) also include channels of sufficient depth and width to be seismically detectable. If we do not see sand bodies like these seismically, it could only be because of inadequate contrast of acoustic impedance. The contact with the overlying shale may be transitional, but probably not over a depth range which would weaken the reflection at normal frequencies; the channel-base reflection should be fairly abrupt.

(2) Bar-finger sands occur only in birdfoot deltas, of which the obvious example is the Mississippi. The depositional mechanism is quite different in this case; the sands are transported down the open distributary, deposited in a distributary-mouth bar at the open sea, and then reworked and transported to form bars along the sides of the distributary—like a sock turned inside-out. In plan, bar-finger sands are linear, or bifurcated seaward. Figure 3.6–4 is taken from Gould (1970, after Fisk (1961)).

Seismically, ancient equivalents of the Mississippi bar-finger sands should be detectable—again provided that the combination of porosity and saturant gives an adequate contrast of acoustic impedance. The lenticular cross-section means we shall never see the edges, but these are transitional anyway; all we shall see is the central thick zone as an elongate amplitude anomaly, without diffractions.

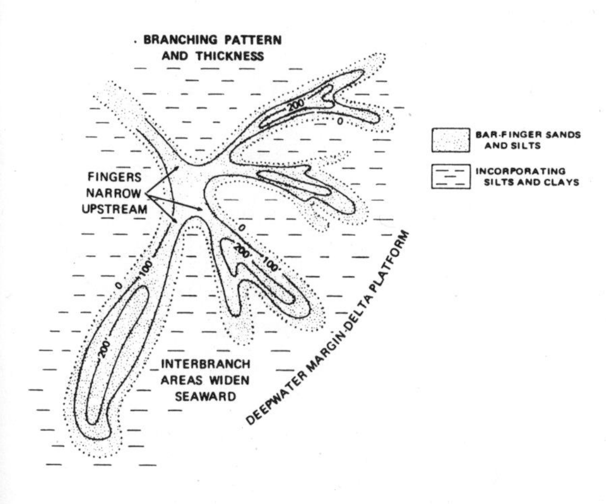

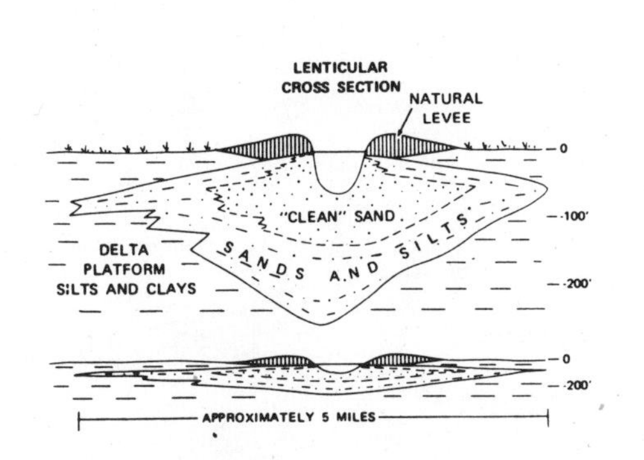

FIGURE 3.6–4
(After Gould, after Fisk)

* Booch rhymes with poke (unless it's a joke).

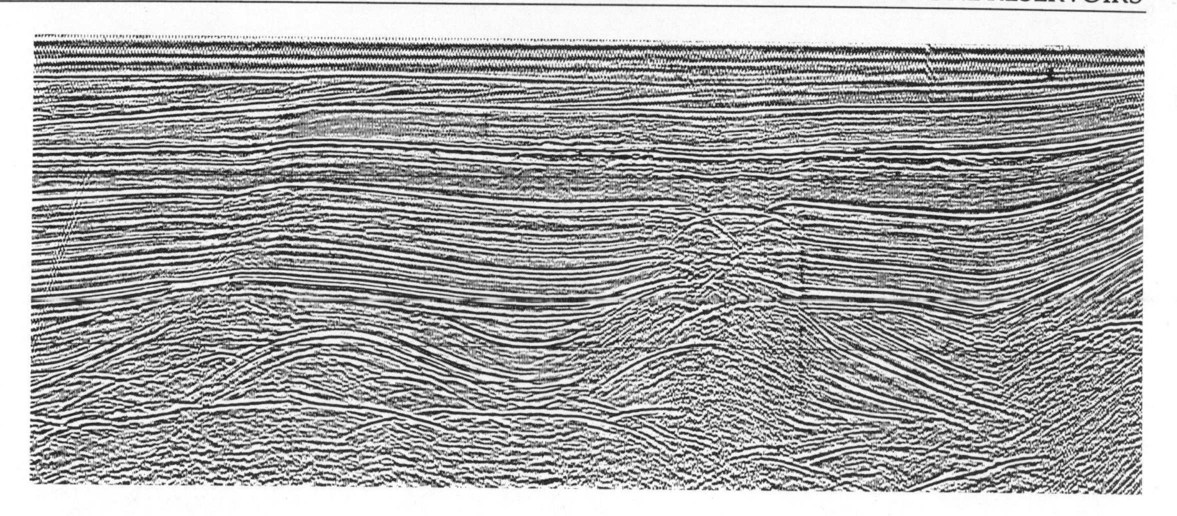

FIGURE 3.6–5 Exercise: *Identify likely locations for deltaic sands. (Courtesy Prakla-Seismos)*

It has been suggested that the mechanism for the generation of bar-finger sands is also likely to generate growth faults across the bar fingers. In this case the thickness of sand is subject to rapid changes along the finger, and the seismic expression may appear as a string of amplitude surges similarly aligned.

(3) A distinct component in the generation of delta-front sheet sands is the building of barrier islands at the seaward margin of an abandoned delta, or along the coast from active deltas. These islands are convex seaward, and are formed by the combined reworking and sorting action of waves, currents and tides.

If these delta-margin islands show seismically, it is as local thickening of the delta-front sheet sands. These changes of thickness may produce increases of reflection

FIGURE 3.6–6 Exercise: *Comment on types and significance of reflection alignments.*

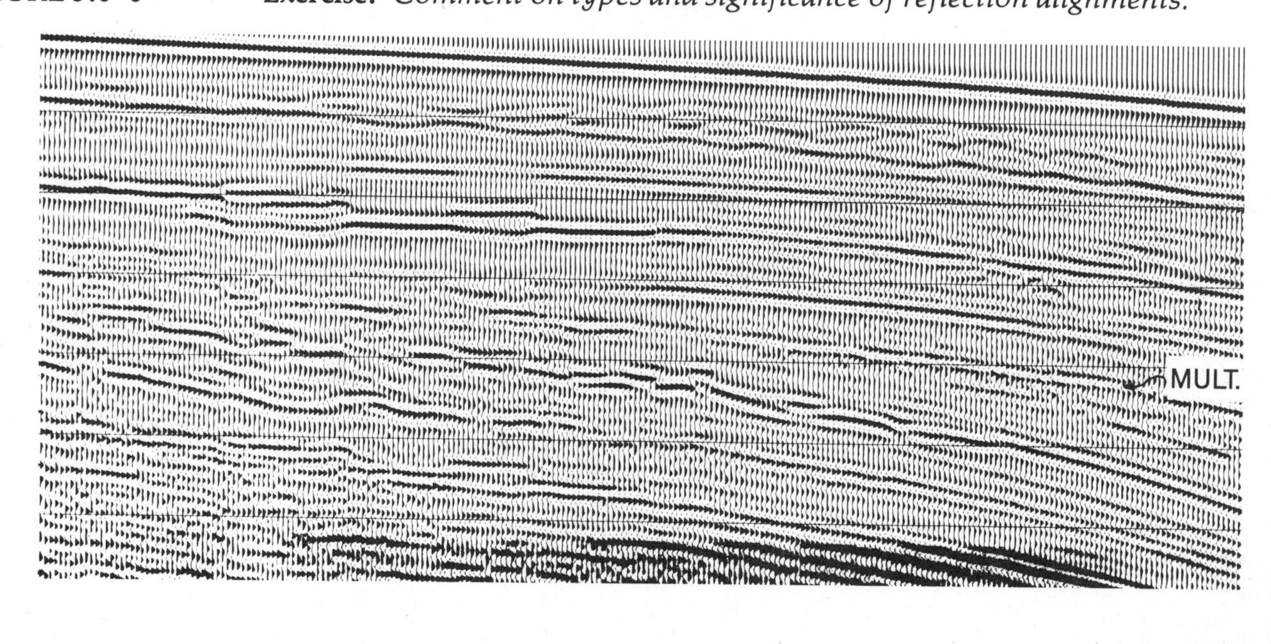

amplitude in themselves; superimposed on these increases may be increases (gas) or decreases (oil) associated with the improved porosity produced by the reworking of the sea.

So where are we now? Clearly the only workable strategy, in the light of all these considerations, is to start with the oblique-progradational signature of a delta, and to work from there. Our first check must be on subsequent regional tilting, because basinward tilting may decrease the appeal of reservoirs at the delta margin; on page 63, for example, the most easily specified target may be the updip limit of the delta-front sheet sand. But if conditions encourage us to look at the body and seaward limit of the delta, we start looking for amplitude and character anomalies on the deltaic plain. There is no doubt that such anomalies are observed on real sections. Then we map these anomalies in plan, in hopes that a pattern will emerge—of meanders, or distributary channels, or barrier islands. Probably it will not, because of the complication of channels and bars stacked one on top of the other. Even so, it is likely that an area showing a profusion of such features will have porosity better than that of the delta-front sheet sands as a whole.

3.7 Marginal-marine sand bodies

Geologically, there are several distinct mechanisms leading to the formation of individual shoreline sands. Seismically, however, the characteristics of these sands are much the same, and here we recognize only two classes: beach and bar sediments, and strike-valley sands.

(1) The first class, then, is the beaches and bars. In sections 3.2.1 and 3.2.2 we discussed transgressive and regressive sheet sands, and noted that a sheet of uniform thickness requires a fortuitous balance between sediment supply and rate-of-change of relative sea level. It is unusual for such a balance to be maintained for long (although well-documented and extensive sheet sands do occur in the foothills of the Rockies, in the Gulf Coast and in the Appalachian Basin). More likely, changes in these variables leave pronounced sand bodies in some places, with zones of non-deposition between them. Even when the sand deposition is sufficient to give good thickness everywhere, the sand is likely to be

FIGURE 3.7-1
(After Selley, 1976)

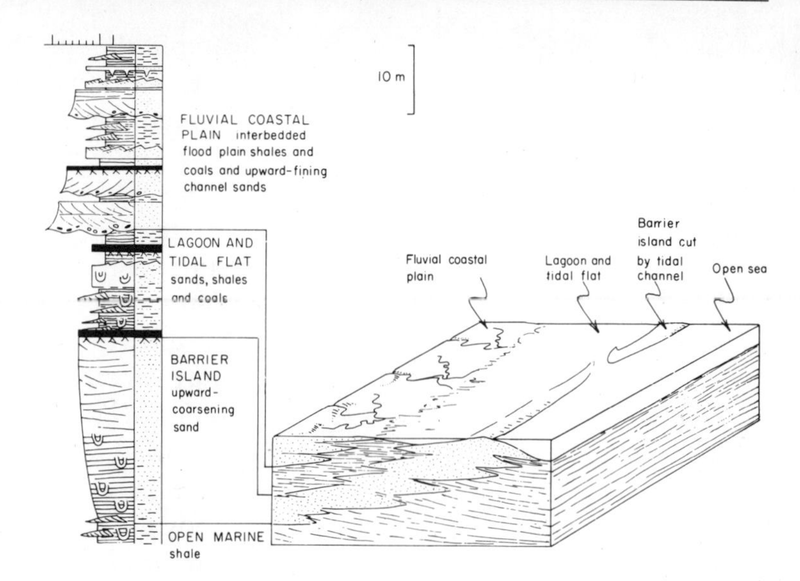

in a succession of individual units. Thus although figure 3.7.1-1 illustrates a continuous sand body like the regressive sheet sand of section 3.2.2, it is more likely that both regressive and transgressive units are built up as shown in figure 3.7.1-2 (regression followed by transgression, in a period of overall rise in relative sea level).

The present class, then, includes offshore bars, barrier bars, baymouth bars, tidal-flat sands and beaches. Seismically, the important characteristics of all these sand bodies are:

- They are lenticular in section.
- They are generally parallel to the shoreline.
- They may be multiple, in stair-step fashion, commonly with shale breaks between them where they overlap.
- Some of them tend to be relatively straight on the seaward side and irregular on the landward side.
- They are always elongate, with the width being up to several kilometres and the length often 10 times the width.
- The thickness is controlled by wave-base considerations, and seldom exceeds 12 m (40 ft).
- Usually one of the vertical boundaries is sharp, the other transitional to some degree.

We can see immediately that these bodies are difficult, in a seismic sense. The lenticular section means no diffractions. The narrow width means that only the big ones, and the ones at shallow depth, are free of weakening effects associated with the diameter of the reflection

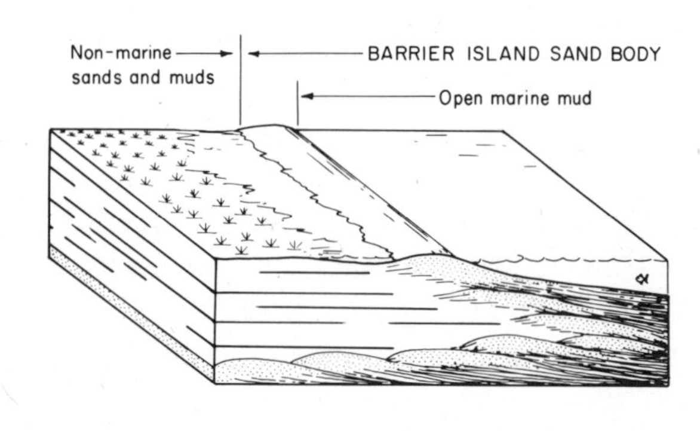

FIGURE 3.7-2 *(After Selley, 1976)*

zone. The stair-step relation implies complicated interference patterns. And the small thickness militates against both detection and confident interpretation.

As so often before, our strongest tool is the recognition of the environment. In particular, we are looking for shoreline indicators; these may be, for example, onlap at an old land surface, or changes in the pattern of this onlap which suggest an intermediate regression. Other sea-level indicators which we can use are deltaic plains evident on other lines, and the characteristic hummocky appearance of shallow-water deposition. In some cases, the limestone stringers of section 3.2.1 are also of value, in indicating stillstands.

When the sea level at a certain time is fixed on one section, we look at the equivalent place on parallel lines, and hope to see a reflection character (both in a vertical and horizontal sense) which is expressed on several lines in a manner suggestive of a shoreline sand body. Again the best chance of a definite identification corresponds to a porous and gas-saturated sand, so that we are, in effect, correlating a bright spot from line to line. In many other cases we shall see only a zone of variable amplitudes and broken continuity, whose position and general character correlates from line to line at the same relative sea level.

Figure 3.7–3 illustrates three pulses of regressive sand growth; the sand bodies are delineated satisfactorily at a depth of 6 km (20,000 ft) even though they are only about 1-2 km wide.

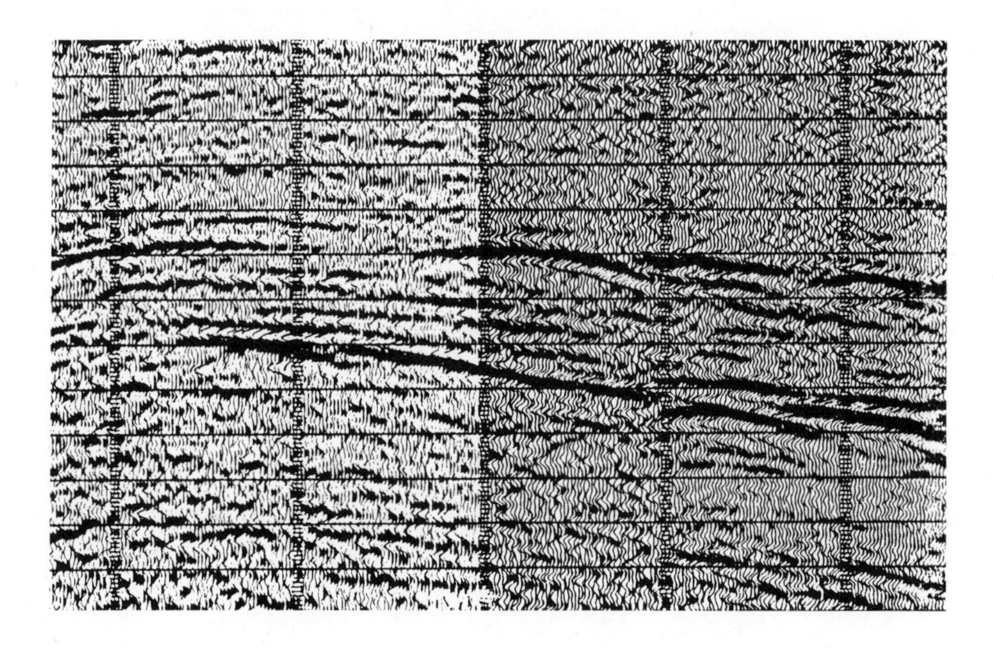

FIGURE 3.7–3

One of the characteristics of such bodies, as we noted above, is the probable existence of shale breaks between them. In the case of figure 3.7–3 we can infer where the shale breaks are likely to be; often, of course, we cannot. This means that one piece of information useful for economic analysis before drilling is not available to us. However, we do better with another piece—the net sand thickness. Figure 3.7–4 reminds us once more that the composite seismic amplitude from a thin sand is linked surprisingly closely to the **net** sand thickness. Intuitively, would we have guessed it would be as good as that?

We have agreed that transgressive and regressive sheet sands require an unusual balance between sediment supply and rate-of-change of relative sea level. Also unusual, but possible, is a prolonged balance which keeps the shoreline substantially **stationary** during a period of large sediment influx; the result can be a most excellent reservoir. The sand can be very thick (1500 m or 5000 ft in the Frio barrier-bar of the Gulf Coast), and therefore it, or parts of it, can be suitable for the type of seismic analysis set out in section 3.1. However, the sand remains a shoreline sand, being long and narrow. The base may be transitional or it may be an unconformity; the top depends on the circumstances

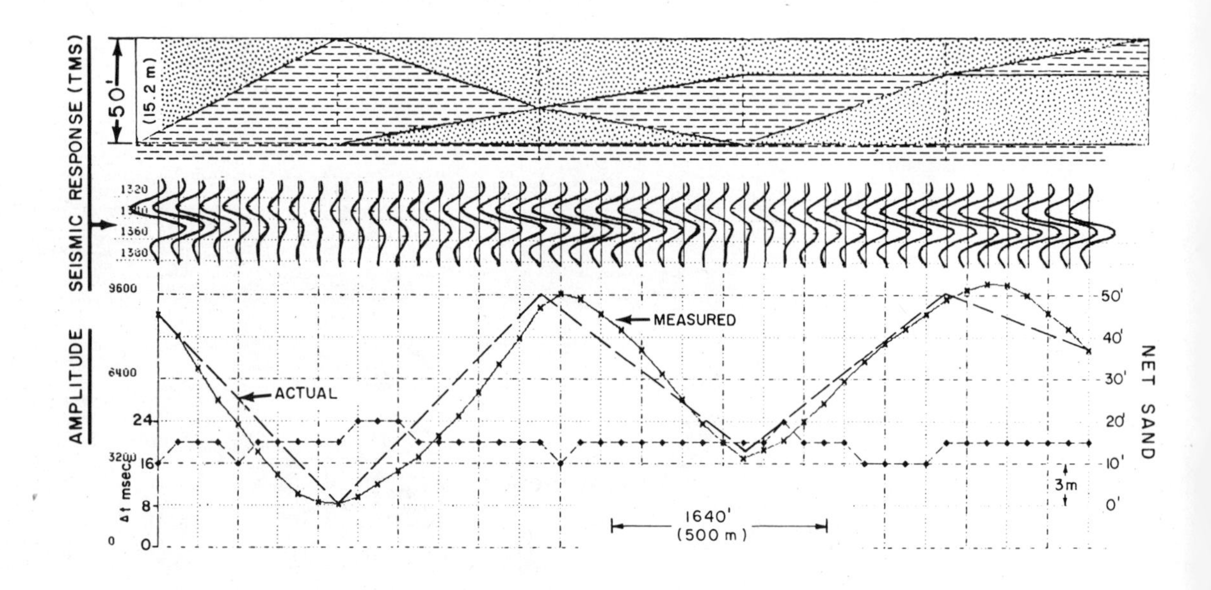

FIGURE 3.7–4 Effect of shale thickness and distribution in a thin sandstone on seismic reflection from sandstone unit. Seismic wave shape is invariant, but seismic amplitude is directly correlated with net sandstone thickness. *(After Schramm et al., 1977)*

of the final regression or transgression, and may be transitional. The lateral contacts (into non-marine deposits on the landward side and marine shale on the seaward side) are transitional. We therefore expect a local reflection anomaly (or stack of reflection anomalies), without diffractions, of an areal form corresponding to a strand line. We might well fail to recognize it; the clue would be there, however, if an obvious onlap succession appears to cease without evidence of a sea-level fall from the highstand, or a subsequent regression in a continuing rise. We would also look for compactional drape over the sandstone.

FIGURE 3.7–5
Exercise: *Shoreline sand bodies, please? (Courtesy Prakla-Seismos)*

(2) The **strike-valley** sand is distinctive, from our point of view, in that additional seismic criteria are available for its recognition. Not that we can guarantee to see the sand itself as a definable event—the thickness may be inadequate for that, and the underlying unconformity brings further interference complications. But there is at least half a chance that we can recognize the circumstances in which strike-valley sands are likely to form.

Basically, we are looking for strike-valley sandstones wherever the reflection alignments suggest **onlap over truncation** (SITPA 36); figures 3.7-6 and 7 illustrate the situation.

At the unconformity surface we expect to see all the characteristics of an angular unconformity, as set out in section 2.7. But now we add to these the evidence of a transgressive sea, with shoreline effects present in the zone of onlap. This combination of characteristics is fairly distinctive; the shale grain is usually quite clear near the shoreline, and the hardness variations below the unconformity (which are basic to the generation of strike-valleys) are virtually certain to show as reflections.

Our interest is increased if the erosional ridges are approximately parallel to the shoreline trend—which also is probably determinable if we have a grid of seismic data. In this case we hope for several strike-valleys, and for a consequent-stream valley joining them to the shoreline; all have the potential to contain sand reservoirs.

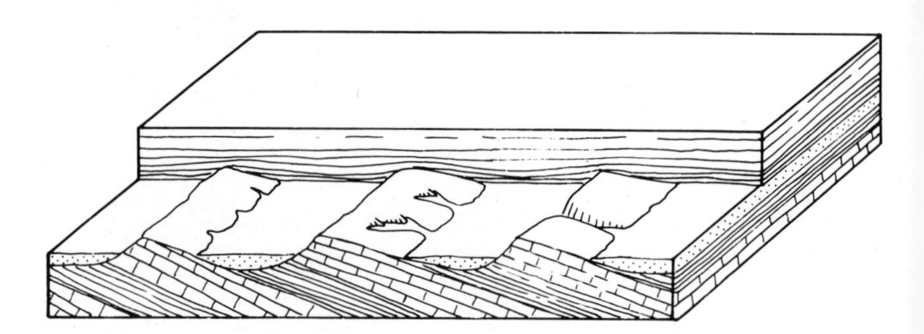

FIGURE 3.7–6 *(Selley, 1976)*

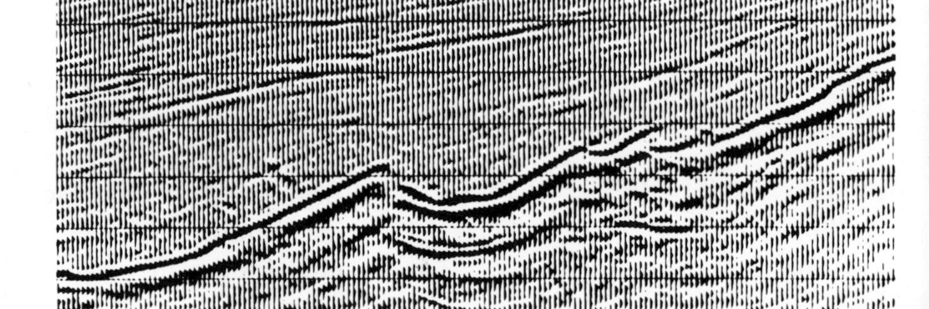

FIGURE 3.7–7
(Courtesy Seiscom Delta)

Exercise: Yes, or no?

Strike-valley sandstones may occur in rings around a diapiric uplift. For example, an incipient salt dome may cause an uplift (with subsequent erosion) during a period of lowstand; a later inundation, as the relative sea level rises, is likely to leave sands in the annular strike-valleys (figure 3.7–5).

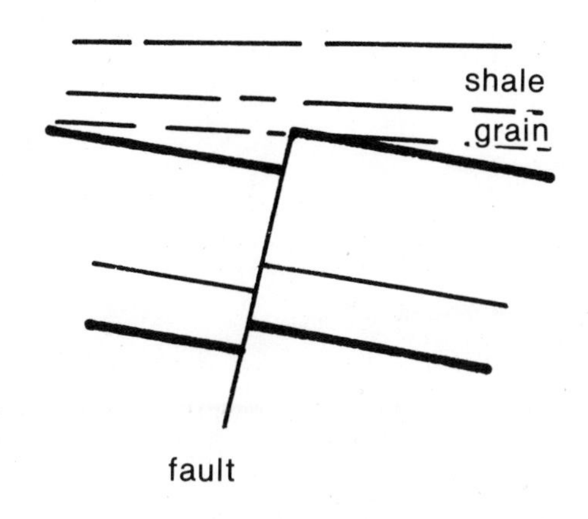

It is also possible to generate strike-valley sands at a fault scarp. Seismically this is more complicated, but still worth a search. We are looking for the fault-scarp signature (as discussed in section 3.5.1) buried under a clear marine transgression—again with a preference for conditions where the shoreline trend is approximately parallel to the fault scarp.

When we come to propose a drilling location on a strike-valley sand, we remember the cautions of section 3.3; careful migration is essential if we are not to miss the sand.

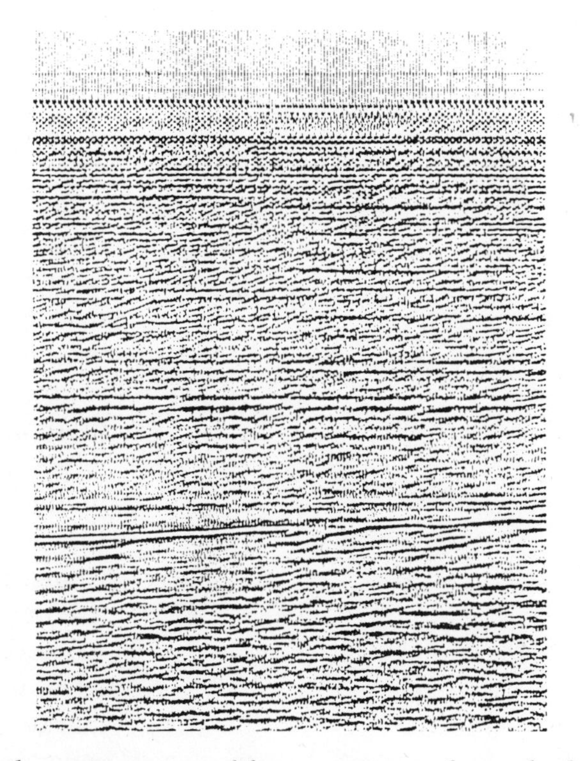

FIGURE 3.7–8 NOTE: This section is a rightward continuation of figure 3.2.2–3; from the latter it is clear that we are indeed dealing with an unconformity. *(Courtesy Amoco)*

Exercise: *Probable location for strike-valley sands?*

Routes of sediment dispersal from river mouth, through longshore currents into submarine canyon by gravity flows and turbidity currents to submarine fan on basin floor. Solid arrows show system of sand dispersal whereas dotted arrows show mud dispersal. *(From Moore, 1972; reprinted by permission of the Geological Society of America.)*

FIGURE 3.8–1 *(From Klein, 1975)*

3.8 Submarine fans and turbidites

Figures 3.8–1, -2 and -3 give us the concept, the three-dimensional view, and the profile view. The first requirement, obviously, is that we should know we are in deep water (figure 3.8–1). If we have the luxury of extensive seismic coverage, we hope to see the sediment source, the shoreline, the sigmoid appearance of a prograding shelf in deep water, and the downlap at the base which gives us the final assurance. If we have only limited seismic coverage in the area of search, we look for a shale grain exhibiting sheet drape over old topographic highs, and take that as our indication that this deposition occurred in deep water (SITPA 417-419).

As we climb the continental slope toward the shelf edge, we may notice that the shale grain becomes more marked; still there may be nothing we would call a reflection, but the general amplitude and frequency of the grain may increase. This suggests that the shale now includes several or many thin layers of deep-sea sandstone, each too thin to show as a reflection; these result

FIGURE 3.8–2 *(After Selley, 1976)*

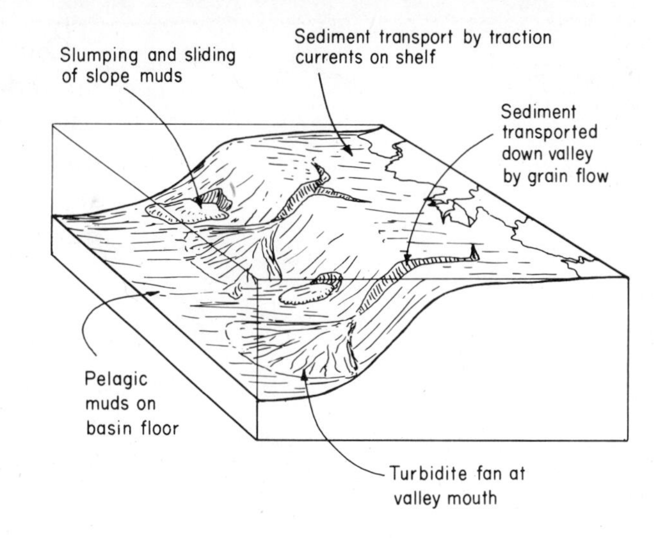

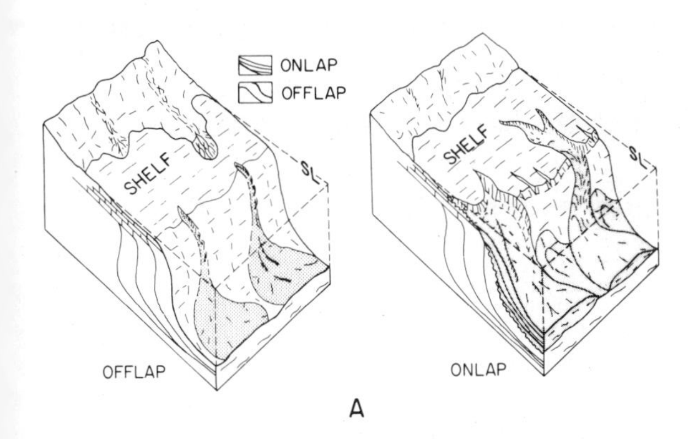

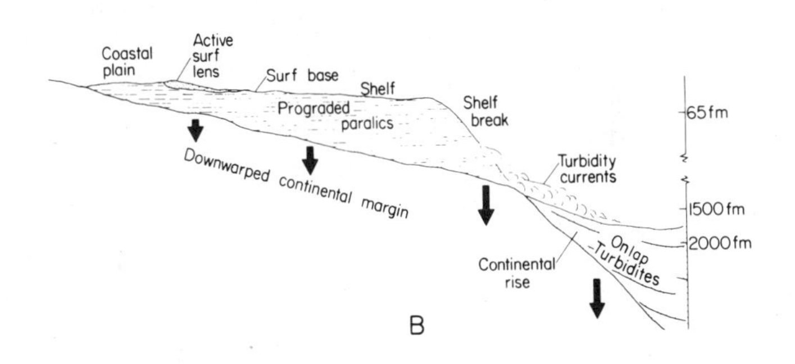

The nature of offlap and onlap deposition—two fundamental depositional styles that characterize many slope systems. **A.** Block diagrams that illustrate general processes: offlap occurs during sustained sediment supply provided by deltas, fan deltas, and highly productive shelf-edge carbonate environments. Onlap conversely occurs when sediment supply diminishes and erosional processes rework shelf or paralic sediments, commonly via submarine canyons. Offlap reflections define the basinward progradation of slope deposits and onlap reflections mark periods of landward recession of slope depocenters. **B.** Schematic representation of onlap processes. *(After Dietz, 1963.)*

FIGURE 3.8–3
(From Brown & Fisher, SS)

from sudden and widespread releases of sand from some cataclysm at the shelf edge. Close to the base of the shelf edge, these widespread sheet sands may become thicker; in particular they may pile up against an irregularity on the sea floor to create a prospective reservoir body. Such accumulations may show as mounded shapes full of chaotic reflections, or merely as a blank break in the continuity of the deep-water grain.*

* In the literature it is said that the seismic signature of a turbidite deposit is an "acoustic void." Unfortunately this term is used equivocally; sometimes it is used to mean the result of substantial absorption (as in the gas-charged shallow sediments of parts of the Caribbean, which allow no reflection to be observed below them), and sometimes it is used to describe an interval which returns no reflections because it contains no acoustic contrasts—but which does not impede the recording of reflections from below. In face of this ambiguity, the term acoustic void is better not used.

Our problem, of course, is not just to locate the mounds; seismic sections are full of evidences of shelf-edge slumping and deep-sea mounds. The problem is to make a judgement as to which of them may contain significant sand. In the case of the isolated deep-sea mounds, this is a difficult judgement to make from the seismic data; all we can say is that it is probably desirable to have significant acoustic contrasts within the mound. We are not too much interested, therefore, in the steady onlap filling of sea-floor lows, and we would certainly hesitate before asking for reservoir properties in units which show no reflections at all.

So mounds showing evidence of deposition in deep water—and preferably showing discontinuous or chaotic internal reflections—become exploration targets. And if there are diffractions too, from within the mound, so much the better. It is no promise of sand (the inside of an overpressured shale mass can look much the same after movement), but it may be all we have.

Of the many examples of mounded turbidite reservoirs, we might select one of the gas- and oil-fields of the Tertiary basin in the central North Sea. This basin is rich with turbidites derived from slumping off the Shetland platform; figure 3.8-4 shows (at 1.8-1.9s) the turbidite reservoir of the prolific Frigg field. Would we have drilled this (the largest offshore gas field in the world) in an untested province and in the absence of the structural play below?

Yes...if we had recognized the mound as the cross section of a submarine fan—by its shape (as in the alluvial-fan example of figure 3.5.1-1) and by its position at the foot of a recognizable submarine canyon (SS 234-237 and figure 3.8.2). The submarine canyon is distinguished from a stream-cut channel by a semi-circular cross-section. The canyon represents a *river of sand* disgorging itself into the fan; the prime target is then the massive and braided sandstones of the upper part of the suprafan-lobe region (figure 3.8-5; Walker, 1978; Siemers, 1979—see 2.86s on figure 3.2.3-2).

Seismically, submarine fans seem very reasonable objectives. The dimensions are often adequate, and sometimes much more than we require; examples in the literature include some where the fan as a whole is 100 km across, where the upper fan region is itself 20-30 km, and where the thickness of sand is 100 m or more (Walker, 1978; Normack, 1978). Clearly the search for such targets should be part of **standard** seismic practice.

FIGURE 3.8–4 *(Courtesy Seiscom Delta)*

FIGURE 3.8–5 *(After Walker, 1978)*

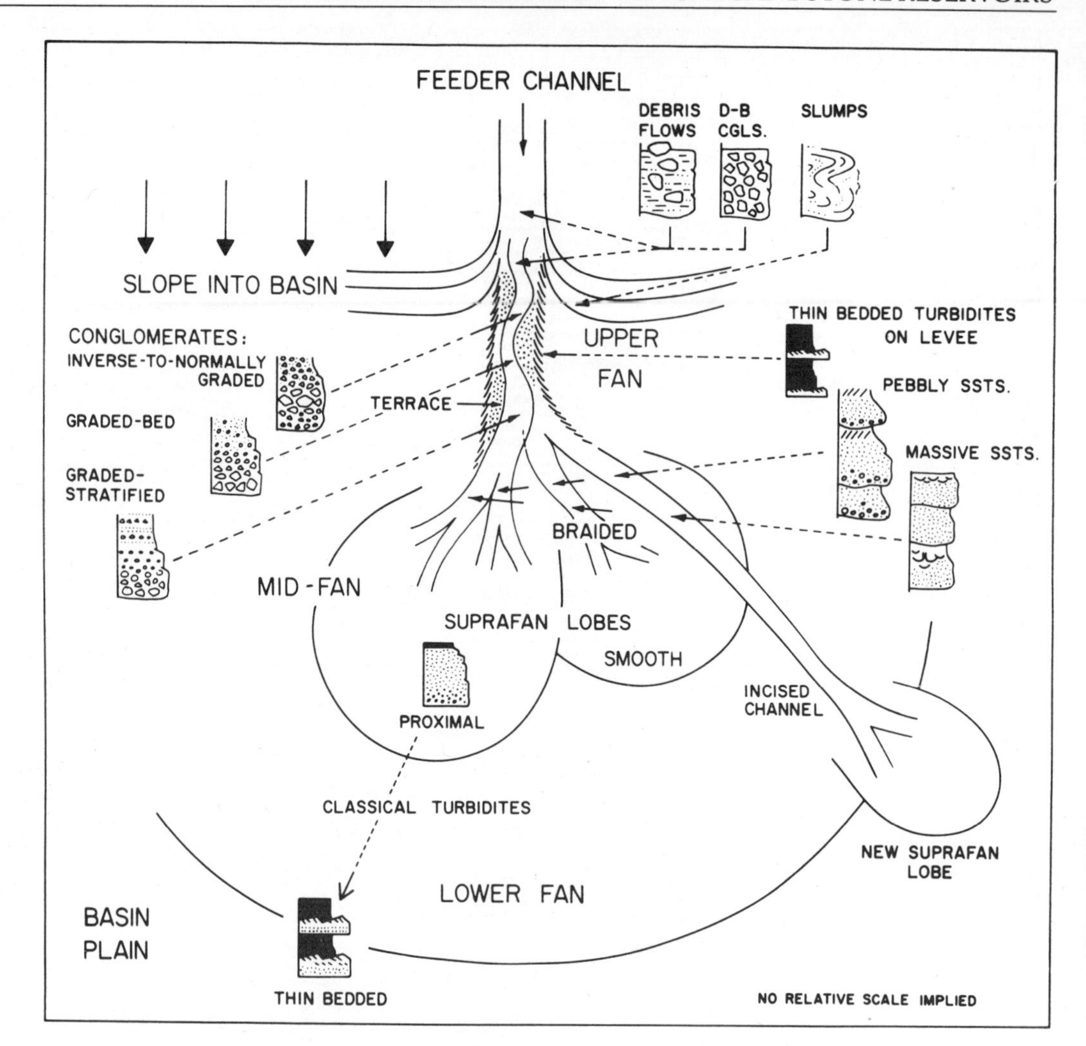

FIGURE 3.8–6 **Exercise:** *A likely location for a submarine fan? (Courtesy Prakla-Seismos)*

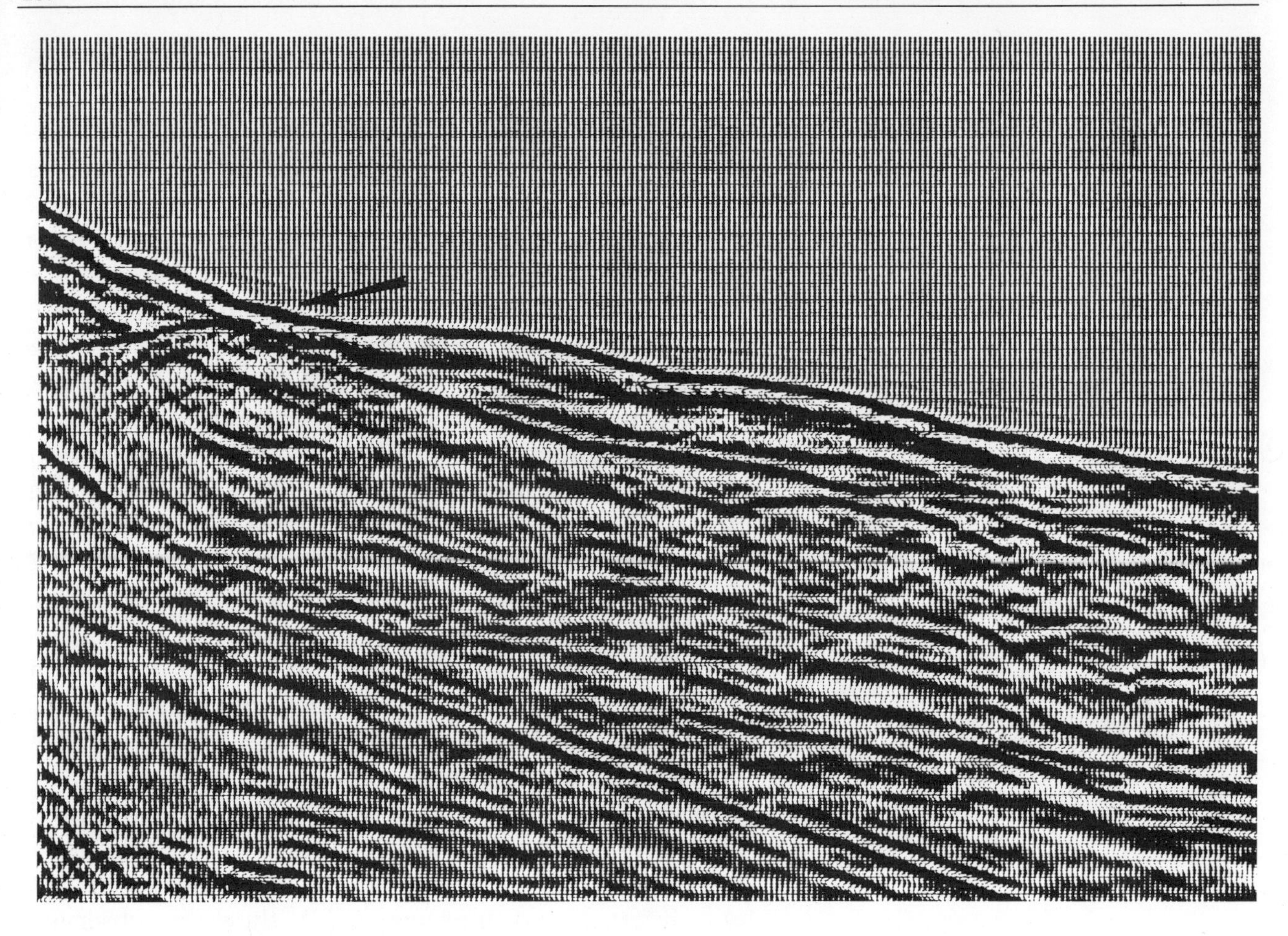

FIGURE 3.8–7 Exercise: *Where are the massive sandstones? What is the arrowed feature? (Courtesy USGS)*

3.9 Some cautions

In introducing this chapter, on the seismic signatures of typical sandstone reservoirs, we reminded ourselves that our first concern had to be with the **depositional** signature. Thereafter, we recognized, we would have to ask about the subsequent geological history, and about the contributions to the final signature which are specifically seismic rather than geological. So how do we stand now?

First, we have to accept the **diversity** of real geology. To learn geology, and to think about it, we must find **order** in it; this leads us to fairly simple models of the type we have been discussing. In a gross sense, these are undoubtedly sound, but every new outcrop we study—and every new seismic section—is likely to contain some feature which does not conform neatly to a model. No two deltas are identical, for example; each one's form depends on a hundred different agencies acting in concert, and something is always different. So we must not become too model-bound.

Second, we must recognize that the seismic method **blurs** everything, and blurs it in three dimensions. Two features which are spaced apart and geologically distinct will interact and interfere when they are observed seismically. Multiple gas sands stacked one above the other, delightfully simple in form and separate in position, may have a seismic expression which is extremely confused. A sand mound, clearly expressed in outline, may have a continuous reflection appearing to pass through it—because the location of the seismic line is such that the reflection zone covers part of the mound and part of the reflection butting against it.

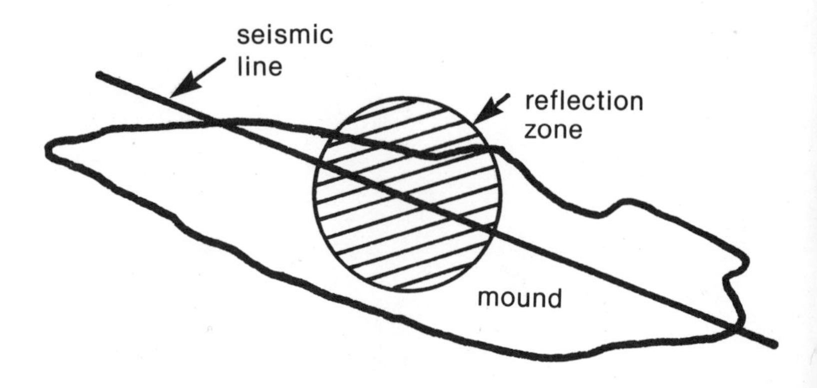

Third, we must recognize the problems introduced by subsequent tilting, subsidence, folding and faulting. The tilting, as we have already discussed, causes any gas present to occupy an area which does not represent the form of the depositional reservoir; this may distort the seismic signature beyond recognition. The subsidence may provide the tilting action; it may also cause drape over sandstone bodies, and distortion of thickness measurements by lateral changes of velocity. The folding may obscure the interrelationships existing at the time of deposition—particularly in the case of apparent growth structures; it also distorts the seismic picture as regards both the amplitudes and spatial positions of reflections. And the faulting just screws up everything.

No one would pretend we have perfect answers (or even adequate answers) to these problems. But some steps are clear:

- Often the visualization of a depositional system demands the **datumization** of a key reflection—the restoration of a depositional surface to the attitude it once occupied (SITPA 444). Thus deltaic plains, shallow-water hummocky sediments and most shelf-type sediments can be restored to the horizontal, thus allowing easier judgements about the degree of subsidence and tilting introduced later.
- The well-established practice of contouring the time interval between two marker reflections often serves the same function. Insofar as the process can be done mechanically, however, there is a balance of advantage and risk; advantage, in the time saved and the objective detachment, and risk, in that judgements may be made on a map which is too far removed from the indications of depositional character evident on the sections themselves. And we have agreed that machine contouring is not generally allowable in channel situations.
- If there is any risk of interval velocities which change laterally (and when is there not, in real geology?), we must make a check of the interval velocities within the contoured interval. Interval velocities today can be computed with surprising accuracy—provided that the data are fair, and that the velocity analysis is done with **great care** (SITPA 285-307).

- Valid migration is essential in many situations—even if the dips are small. **Valid** migration really means three-dimensional migration; if this is not feasible, it it derelict to commission two-dimensional migration without thinking carefully about the effect of the third dimension (SITPA 503-525).
- Structural modelling is valuable as the final validation of an interpretation, and also as an intermediate step within the interpretation if migration and velocity problems are intense (SITPA 448-466).
- Recognition of the depositional environment sometimes requires the restoration of faulted horizons to their unfaulted condition. In practice this is often difficult. It is made even more difficult by the fact that velocities through faulted zones are modified by fracturing and recementing—and in any case are not well measured seismically.
- The detailed resolution of faults requires many more seismic lines than are necessary for the definition of structure. Even more are required for the tracing of channel-fill reservoirs, shoreline sands, bar-finger sands, fans and other features whose recognition depends on **shape**.

As we agreed above, no one would pretend that we have perfect answers.

Of the several thought-provoking conclusions evident from this course, one is particularly disturbing: *There must be thousands of potential reservoirs which have gone unnoticed on seismic sections, and there must be thousands more which just do not show at all on seismic sections.*

4 The Place of the Seismic Method In Exploration For Sandstone Reservoirs

4.1 Seismic stratigraphy

SITPA was written before the publication of AAPG Memoir 26 on seismic stratigraphy. SITPA therefore attempted to give a digest of seismic stratigraphy to the extent that its principles were published at the time (SITPA 413-445). Now that Memoir 26 is published, however, it is clearly inadequate to recommend anything less than study of that volume in its entirety; the book becomes required reading of every geologist and geophysicist. Surely it is the most important work in our field today.

Historically, it has been generally indefensible to go out in a virgin area—other than bright-spot country—to look for stratigraphic sandstone reservoirs with the reflection seismograph. The problem always was—Where to start? Today we see the answer.

We start with the recognition, on our sections, of depositional environment.

The best example, clearly, is the delta. If we are fortunate enough to see on our sections the characteristic oblique-progradational signature of a delta, we are led naturally to a number of potential reservoirs.

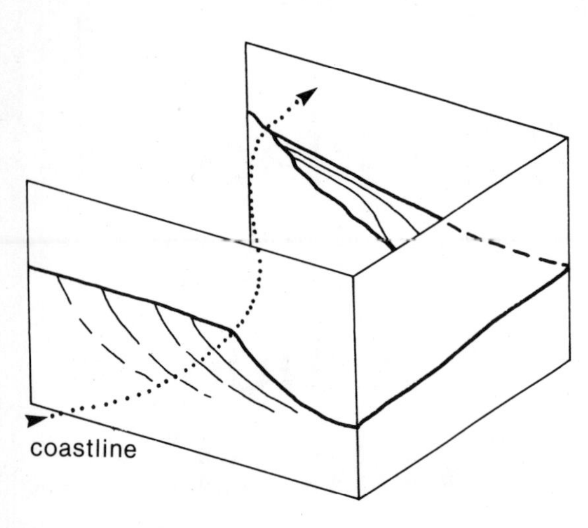

FIGURE 4.1–1

First, having selected a stage in the development of the delta, we attempt to find the position of the time-corresponding shoreline on other seismic sections (figure 4.1–1). To do this we follow the appropriate delta-front reflection (here justifiably assumed to be a time-stratigraphic horizon) down into the section, along on a strike line, and back up on the next dip line. There we hope to find it onlapping in a reflection complex whose nature suggests shoreline conditions.

Repetition of this step on other lines allows us to determine the probable shoreline at the time. Perhaps it becomes clear that the delta has a pronounced lobate form; then we infer that the delta is more likely to be dominated by the river than by longshore currents. This might lead us to concentrate on the delta itself, rather than on barrier and other shoreline deposits whose thick formation depends on long-shore currents. So we would start the detailed search—looking for local thickening of the delta-front sheet sand, and for bar-finger sands, and for distributary-channel fills. Then we might look inland for the point bars of a meandering river, confining our search to the region generally perpendicular to the basic shoreline trend.

In all of this, we would be very conscious of the likelihood that these features of the deltaic plain are underlain and overlain by similar features corresponding to earlier and later positions of the delta. We accept that these complications may be impossible to unravel. We accept even that the seismic expressions of the features we seek may be totally uninterpretable in themselves; it is only their position relative to the delta which sets them apart from a thousand other minor anomalies on the section.

If we have subsequent favourable structure as well, then obviously so much the better. But we may need to remove the structure by datumizing—or in some way display the interesting anomalies riding on the structure—before the depositional origin of the local reflection anomalies becomes clear.

And we are going to drill on reflection anomalies which are strong, or on those which are weak? All we can do is to search for an anomaly representing a sand body sufficiently thick and sufficiently isolated to allow us to estimate the polarity of one of the bounding reflections. For channels and point bars we would prefer the bottom-sand reflection; it is more likely to be a sharp discontinuity, and we hope to be safe in assuming that the underlying material is a marine shale. If the bottom-sand reflection is positive, we are hoping for gas and we drill on the strong amplitudes. If it is negative we have an agonizing choice; strong amplitudes are likely to mean low porosity in a sand, while weak amplitudes could mean either good porosity in a sand or some intermediate and unappealing facies.

At this level, of course, we are prepared to be wrong sufficiently often to keep us humble.

If the sections show evidence of a subsequent marine transgression over our delta, we would naturally be led to look for later deltas further inland, using the delta-prospecting strategy of Busch (1974, pp. 140-144).

Or perhaps the shape of our delta is not lobate, and some of our attention switches to beaches and bars along the shoreline. Then we hope to determine, by the number and intensity of reflection anomalies, in which direction from the delta the longshore currents were building the sand bodies. Again we accept that the reflection anomalies may be totally uninterpretable in themselves; it is only their position—along a shoreline receiving vast quantities of sediment—which excites our interest. Then we must take full advantage of the information which seismic stratigraphy reveals on the position of sea level at different times. Thus we are able to estimate the elevation of land masses, mountain chains and flat plains above the sea level of the time, and so to deduce the likely locations for alluvial fans, channel deposits requiring moderate slope, meander belts, strike-valley sands, marginal-marine deposits and turbidites.

And so on. Each of the sandstone bodies we discussed in the last section is associated with a depositional environment, and the techniques of seismic stratigraphy provide a means to define these environments. Thereafter many types of sandstone body allow corroboration by their shape, but it is the techniques of seismic stratigraphy which give us the all-important start.

An interesting example of seismic stratigraphy at work is given by Louis and Mermey (1979).

4.2 The extension problem

So seismic stratigraphy gives us new hope for finding sandstone reservoirs not previously within our seismic grasp.

Even in virgin territory.

However, we shall not hold back the independent efforts of those who believe that what it takes is courage, and faith, and optimism, and a commitment to the free enterprise system, and the money to drill 34 dry holes. Because sometimes (1 in 35, for example) they will be lucky. Then we must ask: Given a discovery, what can the seismic method contribute to the location of the next hole? Where is the rest of the field?

This has been called *the extension problem*. We can tackle it thus:

- We now know the depth and thickness of the target; this allows us, as we shall discuss in the next section, to design an optimum field technique.
- If we check-shoot the well, we also know the seismic time of the target.
- If we have cores of the reservoir, we study the characteristics of vertical sections (as set out in the MacKenzie table), and so form a judgement on the depositional environment of the sand body. If the cores are oriented, this in itself may suggest a direction for the next well—but for a sand body laid down in a high-energy depositional environment such a step would be risky.
- From the cuttings and the logs, we know the lithology of the sequence above and below the target. We also know the porosity. We know whether we have gas or oil, and we have some idea of the water-saturation. So we already have some feeling for the type of seismic response we expect—in one dimension.
- At this stage, then, we probably know what the enemy is. We know whether it is thinness of the reservoir, or whether it is a lack of reflective contrast between the reservoir and the bounding materials, or whether it is in the transitional nature of these boundaries. There remains a chance that the enemy could be the narrowness of the reservoir; it could be a shoe-string, at a depth where it intercepts only a small part of the reflection zone.

- We can become more precise, still in one dimension, by constructing a synthetic seismogram from the calibrated sonic and density logs. The considerations to be taken into account are given in SITPA 527-535. In the present case we would probably assume phase-zeroizing in the processing of the field data, and therefore use a zero-phase pulse for the synthetic. By running synthetics with pulses of different bandwidth, we would soon determine what reflected bandwidth is necessary in order to resolve the reservoir. This is very important information; if we feel that the bandwidth required is unrealizable, with the existing surface conditions, we would **abandon the seismic approach** (SITPA 624). Otherwise, the bandwidth information defines the field technique, as we shall see in section 4.3.
- Then we can repeat the synthetic, over the interval of interest and perhaps 200 ms above, making various changes at the reservoir level. For example, we can make the reservoir sand thicker and thinner (at the expense of the overlying or underlying materials, or otherwise as we think geologically appropriate); an example is given in figure 4.2–1. If the saturant is gas, we can try the effect of moving the fluid contact. Then we are expecting that we shall obtain a match between the original synthetic and the field data at the well, but that the match will be with one or other of the modified synthetics as the seismic coverage extends away from the well (SITPA 566-568).
- If we are encouraged by the synthetics, we go to the field. We design the field technique and the processing requirements, and start with a line through the well. If we have no hint of a trend from other evidence, we probably take the line location which promises least surface problems.
- If we are able to shoot just the one line, economically, we make the comparison with the synthetics before proceeding further. Perhaps it is clear that the match, while good at other levels, is poor at the reservoir level; the conclusion is that the sand body is highly variable within the reflection zone (probably on account of thickness variation or internal properties), or that it is narrower than the diameter of the reflection zone, or concave or convex within the reflection zone.

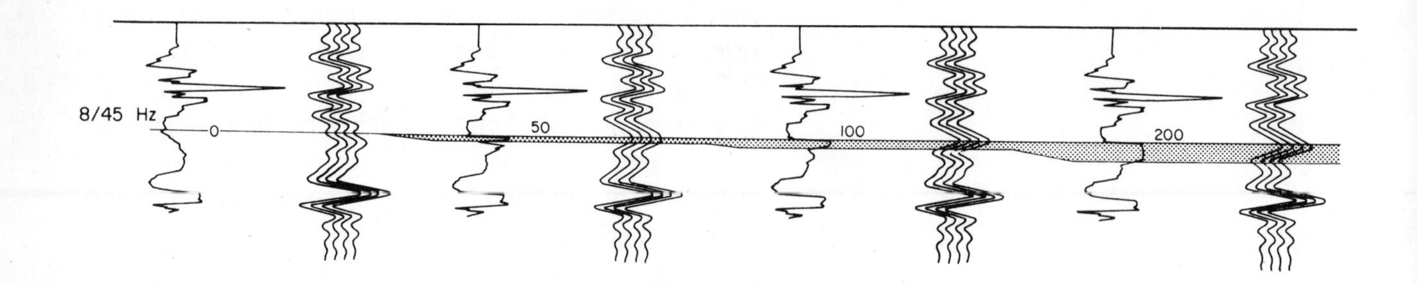

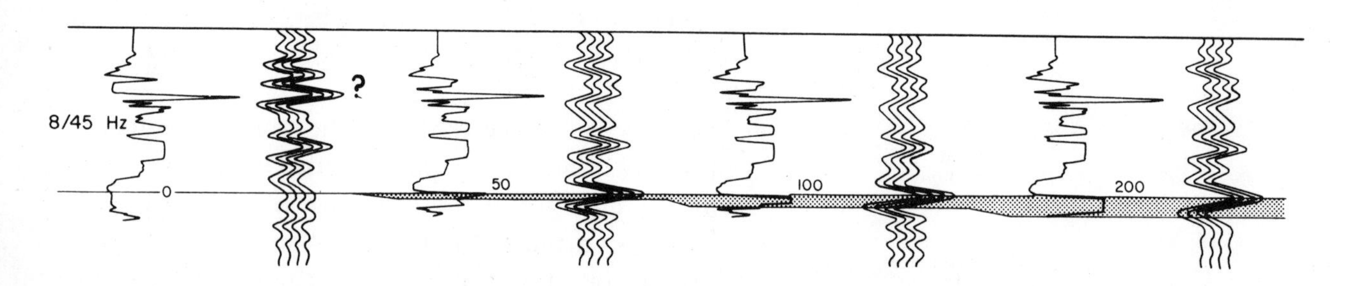

FIGURE 4.2-1 Effect of increasing sandstone thickness on seismic response, lower Wade and Medrano sandstones. Model seismograms are produced by inserting increments of material with appropriate velocity at stratigraphic position of sandstone unit on digitized sonic log. *(After Galloway et al., 1977)*

- But if the match is acceptable at the target level, we then search to see what variations in the thickness of the reservoir or position of the fluid contact are suggested by the matches extending away from the well.
- Under ideal circumstances, we would then return to the field and shoot more lines radially through the well, with a grid lines to tie the ends. (In practice, of course, we may have to do this at the time of shooting the first line.) If everything is well disposed to us, this coverage should tell us where to drill next.
- In this scheme we are assuming that—on the basis of only one well—it would not be economically justifiable to do a major seismic survey directed at

determining the depositional environment in a gross regional sense. So we may not really understand the origin of the reservoir after our local survey round the well, but we hope to learn which way the reservoir goes.

- We must accept that the synthetics cost a little money, and that the local survey costs much more. So someone will have to make the decision whether to do the seismic work, or to drill a step-out well without it. Where drilling is very expensive the argument for the seismic approach is clear-cut, but in other cases we have to admit that the decision is not necessarily easy.
- After all this, we probably find that the oil or gas accumulation is structural anyway. So we all have a good laugh together, and talk a great deal about courage, and faith, and optimism, and...

4.3 Recording and processing considerations

According to our discussions so far, we see two clear functions of seismic data. The first is to resolve the depositional environment—a function which usually requires a regional survey, with long lines and broad coverage. The second is to identify and study a prospective reservoir—a function which usually requires highly detailed coverage of a small area.

Fortunately, the first function—of establishing the depositional environment—can be undertaken with the type of data which is now routinely obtained, and we do not need to say very much about the recording or processing. Probably these three points are all that we need to add:

- Large-scale sections, while sometimes essential for detailed work, are a mistake for broad studies of depositional environment. It is important to be able to see the whole picture in one eyeful. So the vertical scale should be about 5 cm/s (2 in/sec) and the horizontal scale should represent a vertical exaggeration of 5-10 times—a fact which geologists have known since the discovery of fire. With this type of squashed display, the sequence boundaries and the faults (and in particular the transgressive and regressive situations) spring out far more clearly than at normal scale. One caution however: If the scale is smaller, our pencils should be sharper too.

- Judgements based on onlap should properly be made only where there is three-dimensional control (for example, at line intersections). A section at right angles to apparent onlap could show true downlap (SS 58, 121; Roksandic, 1978).
- In general, the agc action in the processing must not be so fierce that we risk failing to recognize the weak reflection grain of a low-energy unit (SITPA 419, 420). However, some method of seeing the very low amplitudes may be important for porous sand reservoirs (as in the discussion of turbidites in section 3.8 above).

For our second function—the identification and study of specific prospective reservoirs—the recording and processing must be appropriate to the problem; standard techniques may or may not be adequate, but in any case the recording and processing must be done with understanding.

Much discussion of the way in which the seismic techniques require selection according to the problem is given in SITPA (620-626, 230, 232-252, 261-265, 278-283, 335, 382-384, 496-501, 549-565, 608-619). Here we shall be content with highlighting a few specific matters:

- Reservoir identification and assessment ordinarily require a much closer grid of lines than is required by a regional study of depositional environment.
- If the problem is resolution of a thin sand, phase-zeroizing (in a wavelet-processing package) is important, but bandwidth is **essential**.
- The limit to bandwidth may be in the earth (SITPA 117-143), or it may have been inserted in the anti-alias filter (for reasons of processing economy), or it may be in the source.
- If it is in the source, we should explore the economics of improving it. For dynamite work, we can usually get better bandwidth by drilling deeper holes and/or using very small charges—many of them, if necessary. For Vibroseis, we can get better bandwidth by extending the sweep and making it non-linear, but we must then spend

much more time sweeping if we are to maintain an adequate ratio of signal to ambient noise at the extended frequencies.

- To the extent that we are successful in extending the bandwidth of the received signal (not just the transmitted signal), we must adjust the anti-alias filter and the sampling interval accordingly.
- To this extent also, we must adjust the length of the geophone arrays and the source array so that we do not lose bandwidth there—at least at target depth.
- In this context it is clearly desirable to use space-and-time-variant arrays. These arrays are realized by dividing the continuous spread into many sub-arrays in the field, and then merging them together on a time-variant and distance-variant basis in the computer. They provide the long arrays desirable during passage of the surface waves, and at the end of the record, while also providing short arrays—with minimum sacrifice of vertical and horizontal resolution—in other parts of the record.
- Since both polarity and vertical resolution are important in many of the applications we have discussed, all processing should have the phase-zeroizing step. But this step is certainly being asked to do more than it should. For example, many interpreters presently believe that a Vibroseis section is more-or-less zero-phase. They remember that the pilot sweep goes through a field amplifier before correlation, and that the vibrators have a phase-lock system, and that the final filters are zero-phase, so there is only the earth to distort the reflection from symmetry. But some Vibroseis crews lock the **velocity** of the baseplate to the pilot sweep (with the expectation that this makes the basic pulse symmetrical); others, perhaps with an eye to the 90° phase shift characteristic of a particle-velocity measurement in the near field, lock the **displacement** of the baseplate to the pilot sweep. In any defined situation, only one of these can be right. Then some processing centres use **minimum**-phase deconvolution without any compensatory steps, and so distort the phase even further. It seems less than ideal to pass the resultant traces through a **statistical** phase-zeroizing process

after all this, when the major phase errors have been introduced knowingly.

- Thus in all seismic work, and for our present concerns in particular, it is better to use **deterministic** compensation for effects we know about, and **statistical** compensation only for the effects we cannot measure directly.
- As was stressed many times in SITPA, the finer points of modern interpretation depend heavily on a new degree of care and understanding in the processing. Amplitudes must be handled intelligently, velocity analyses must be properly positioned and thoughfully picked, sophisticated processing must not be wasted on data which has been sloppily pre-processed, and care should be taken to **interrelate** the processing with the field work (and one processing stage with another) so that the means and the objectives are consistent.
- Most of us, up to now, have subscribed to the view that the final time-variant filter applied to the data is there to give a final improvement to the signal-to-noise ratio, and generally to make the section a little easier to pick. It goes without saying that if our problem is the resolution of a thin bed we wish this filter to be very mild at the top end. It is less obvious, perhaps, that we must be very careful about the low-frequency end. This is because of the sensitivity of transitional reflections to the low frequencies. We have agreed in section 2.5 that the evidence of transitional reflections on the section may be very relevant to the determination of the type of sand body involved; such evidence may be perfectly clear on a section without a low-cut filter. Transitional reflections just disappear when the low frequencies are cut; in this they are generally different from low-frequency events formed by two closely spaced reflections of like sign, which change their character but do not disappear. The distinction can be useful in interpretation.
- Migration is no longer seen as a process reserved for steep-dip country. It is invaluable for clarifying the details of an eroded surface, counteracting in part the blurring effects of the reflection zone, identifying sequence boundaries, revealing small channels and fills, restoring fans to juxtapose

scarp faces where they belong, correctly relating rollover and other traps to the generating growth fault, clarifying turbidite mounds, distinguishing between sand bodies which are truly lenticular and those which appear lenticular because they are fault-bounded, and generally facilitating any judgement which uses the reflection signature of a reservoir in profile or in space.

- Finally we should always be conscious of the importance of display. It is true that oil is found in the minds of men—but not until the basic data have been transferred into those minds; the vehicle for that transfer, in our context, is the seismic display.

For some situations the pseudo-log (or the seismic log) is a valuable display. It has the merit that for layers of appropriate thickness it transforms the seismic display from being one of **contacts** to being one of **layers**. As such it is particularly appealing as a way of expressing seismic results in a way which appears familiar to log analysts.

Of course we have to remember with all forms of display that we never escape from the unwelcome consequences of the seismic pulse and its wiggles—we cannot escape from tuning and interference problems by changing the form of display. One of these consequences is that a thin layer indicated as **soft** on a pseudo-log display will appear to be sandwiched between two materials indicated as **hard**—and the softer the meat the harder the bread. So we have to understand what we are doing—but that is true with any form of display.

Then there is the problem of display for **subtle** features. We have to see, on a grid of lines, some subtle amplitude or character change whose areal extent or shape is suggestive of a certain type of sand body. We have to see it, moreover, when it is superimposed on structure and intercepted by faults. This is not a trivial problem, and it is surprising that it has been given so little attention. To the geologist, the essential input to the definition of a reservoir is a **pair** of maps—the structure map and the sand thickness map. The two messages must be brought together and seen as one three-dimensional reality. To the geophysicist the structure map translates directly, but the sand thickness map translates (in many practical cases) into the amplitude variations observed along profiles. What is needed is an amplitude display which can be seen **riding on the structure**, and which can be seen in three dimensions; the sculpted fence diagram of SITPA 608-619, with the amplitudes quantitatively displayed in colour superposition, is one solution.

4.4 Reflection work with surface sources and a borehole geophone

In section 4.2 we discussed the extension problem—the problem of where to drill next, after a discovery. A related problem—delineation with fewer wells—is discussed in SITPA 566-573. These treatments are supplemented here by more specific reference to the possibilities available with a borehole geophone.

Let us restrict ourselves to the case where there is an acoustic contrast between the reservoir and its bounding rocks, or at a fluid contact. Then the most likely problems we face are the inadequacy of the vertical and horizontal resolution provided by the normal seismic method—the reservoir is too thin, or too narrow, or too faulted.

We can obtain an immediate improvement of the seismic method, in this case, by retaining the surface source but by using a borehole geophone. As the geophone goes closer to the reflector, the high-frequency losses decrease and the diameter of the reflection zone decreases—our vertical and horizontal resolutions improve. Surface-wave problems disappear and, with a good lock-in geophone, signal-to-noise improves.

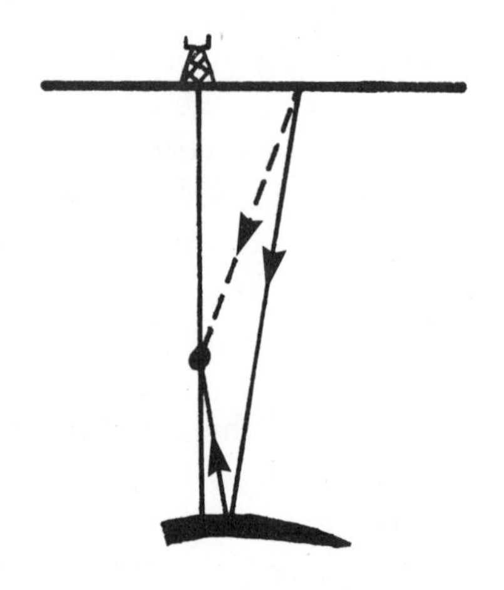

But there is much more than that. We obtain two major advantages from the fact that we have a record of the downgoing signal as well as of the reflected signal.

First we obtain a direct measure of the reflection coefficient of the reservoir boundaries (SITPA 253-259).

Second, we can do what we have always wanted to do: deconvolve the upcoming signal using the downgoing signal. All we have to do (provided we choose the geophone-reflector distance appropriately), is to taper the direct downgoing arrival to exclude reflected ar-

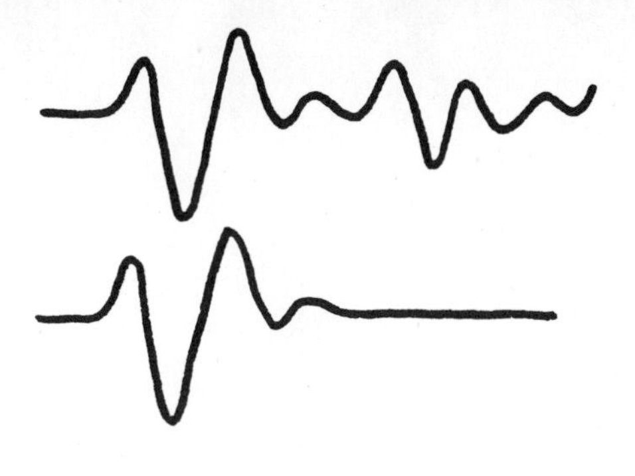

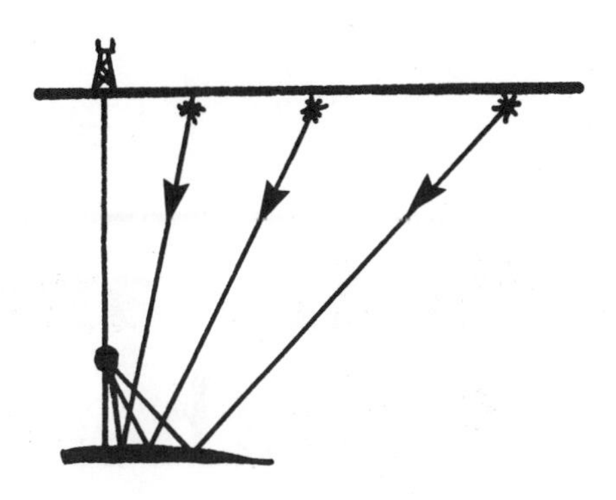

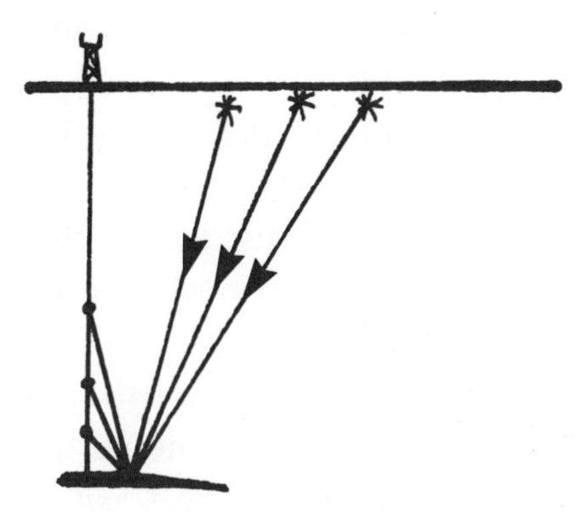

rivals, and then use this pulse shape as the known wavelet in a standard deconvolution program.

These two advantages increase very considerably our ability to measure the detailed characteristics of the reservoir. Then the next stage is to record **profiles** of source positions radially from the well, and so construct sections showing how these detailed characteristics of the reservoir change outwards.

As the profiles extend outwards, the geometry changes from near-normal-incidence to wide-angle, and finally to refraction. The variations in the geophone signal, over this range, contain much additional information on the nature of the reservoir, the saturant, and the fluid contact (SS 40-44; SITPA 111-116).

Finally we may occupy a different geophone position (over a suitable range of depths) for each profile of source positions, and so build up a common-depth-point stack at each of many reflection points over the extent of the reservoir.

There is an interesting feature of this type of stacking. In normal cdp-stacking we assume the geometry and make a search to find the stacking velocity; in borehole cdp-stacking, however, we know the velocity directly—but we must make a search to find the paths with the right geometry.

Now that the contractors are able to provide digital recording and processing of well surveys, and air-gun or surface sources of good bandwidth, these techniques must surely make a significant contribution to the extension problem. They should save many wasted delineation wells. Let's get to it.

5
Summary and Synthesis

(1) There must be thousands and thousands of undetected sandstone reservoirs.

(2) Many of these we shall never detect seismically, because the combination of good porosity and oil saturation can destroy the acoustic distinction between a sandstone and a shale. Thus many such reservoirs return no reflection.

(3) Many of the remainder are uninterpretable in themselves; they are too small on a seismic section, or too subtle, or too much like a host of other (uninteresting) anomalies.

(4) The key to the location of these less obvious reservoirs is the seismic definition of the depositional environment.

(5) Given an asymmetric basin, we ordinarily look for structural and fault-type traps on the steep flank. On the gently-dipping flank, however, we search the hingelines and broad homoclines in hopes of finding **fairways** of multiple stratigraphic traps (King, 1972).

(6) In so doing, we prefer basins exhibiting:
- a record of regional tectonic instability,
- important unconformities and overlaps, and
- alternating marine, paralic and fluvial facies.

(7) In the search for stratigraphic traps, we avoid the complications of large structures (except those generated by diapiric movement).

(8) The key to the seismic definition of depositional environment is the tracing of seismic sequence boundaries, which are unconformities or their correlative conformities.

(9) The transition from unconformity to correlative conformity (particularly in conjunction with thickening sediments) is evidence of the transition from land to sea.

(10) A major objective is to find ancient shorelines—particularly in the areas of and between deltas. We generally prefer the shoreline sediments deposited during periods of overall transgression (MacKenzie, 1972), though we note that individual barrier sands are more likely to be left in place during a prograding phase than during a transgression.

(11) The zone of lateral transition *out of marine shale* is one which is likely to combine source, reservoir and stratigraphic trap.

(12) Deep-water shale is recognized by its weak reflection grain, and sometimes by sheet drape; it is distinguished from low-energy carbonates on the basis of interval velocity.

(13) Every generalization imputing a particular facies to a particular depositional environment must be qualified by the phrase *provided that appropriate sediments were available at the time.*

(14) Very thick blanket sandstones usually originate as aeolian dunes, at the final stage of a regression. Bounding reflections may be discrete, without significant transitions. On the assumption that the saturant is water, the porosity can be calculated from the interval velocity provided that the sand thickness exceeds about 40 ms. If oil is postulated, a range of porosities can be calculated, corresponding to an estimated range of water-saturations. If gas is present, it may show as a strong negative reflection from the top; the fluid contact may also show. In favourable cases, the gas porosity can be calculated from the combination of velocity and amplitude, again for a range of water-saturations.

(15) Thin blanket sandstones may originate as dunes, or as delta-front sheet-sands, or as transgressive or regressive sheet-sands deposited during a rise of relative sea level, or possibly as regressive sheet-sands deposited

during a fall of relative sea level. **Thin** is likely to mean less than 30 m (100 ft). For this category of sheet-sands, and for other sand bodies in the same range of thickness, no explicit calculation of porosity is possible at this stage of the art. However, under favourable conditions of gas-saturation, and if the bounding interfaces are fairly sharp, a gross figure-of-merit (representing the combination of good porosity and good net thickness) is given by calibrating the amplitude of the reflection complex from the sand.

(16) A transgressive sheet-sand usually has a **top** interface which is transitional to marine shale. A delta-front sheet-sand and a regressive sheet-sand usually have **bottom** interfaces which are transitional.

(17) If these transitions occur over depth ranges shorter than a seismic wavelength they have little effect, and the quantitative techniques summarized above remain valid. If they occur over longer depth ranges they act as a high-cut filter; the reflections become low frequency and weak, and quantitative calculations based on amplitudes are jeopardized. However, a thin sand body underlain or overlain by a transition is more obvious, seismically, than the same sand body without the transition.

(18) Thus if the vertical transitions into shale are very gradual, a delta-front sheet-sand and a regressive sheet-sand have a seismic signature which is a fairly sharp reflection with a low frequency tail. Conversely, a trangressive sheet-sand might conceivably have a seismic signature which is a sharp (unconformity) reflection with a low frequency precursor. These distinctions may or may not be possible; certainly they require wavelet processing, intelligently applied.

(19) More useful than these distinctions, generally, are distinctions based on the depositional environment. Basic, in this context, is the seismic recognition of periods of lowstand and periods of highstand. The deposition between lowstand and highstand is characterized seismically by the signature of coastal onlap (which must be distinguished from submarine onlap occurring, for example, in the upbuilding of a submarine fan).

(20) Thus we hope to find transgressive sheet-sands and shoreline sands wherever we see the signature of coastal onlap on an unconformity surface. We hope to find their regressive (or balanced) counterparts wherever we see a

termination of coastal onlap without a fall of relative sea level. And we hope to find delta-front sheet-sands wherever we see the oblique-progradational signature of a delta.

(21) The seismic signature of an unconformity is the angular relationship between reflections above and below, coupled with a reflection of variable amplitude and character (and possibly polarity) from the unconformity surface itself.

(22) Drilling locations on traps formed by the truncation of sheet-sands at an unconformity will be wrong if they are selected without regard for the nature of the seismic reflection pulse and its polarity.

(23) Growth faults require bottom transport of sediments; such sediments are likely (but not certain) to contain sands. Such faults are concave basinward and concave up, and commonly terminate at low angle on the basinward side of shale masses. Excellent structural traps may occur in the rollover zone basinward of the fault; other traps (which may offer multiple targets) occur in the complex system of crestal and antithetic faults generated by the main growth fault. Growth faults also suggest a source of coarse-grained sediments to landward; this may point to a delta or other new target.

(24) Individual sand bodies often make better reservoirs than the widespread sheet-sands, because of their locally superior sorting and porosity. Most important are the fluvial bodies, the shoreline bodies, and the turbidites.

(25) The process of seismic picking, in the presence of the facies variations, porosity variations and saturant variations of the fluvial and shoreline environments, is quite different from normal structural picking. Continuity is no longer a criterion; it may not even be desirable. Reflections are expected to show major changes of amplitude and character, and may even change polarity (particularly if they are weak). The picking **is** the interpretation.

(26) More important than picking is seeing the **shape** of the character changes in plan view, and associating this shape with a type of sandstone body. Thus our concern is to **map** all character changes seen in a favourable depositional environment—even though we may not understand the detailed origin of those changes.

(27) We then relate the indications of depositional environment to the shape of the character anomaly, and try to infer the likely type of sand body.

(28) Thus an alluvial fan may or may not show directly as such; however, we hope to recognize it by the combination of a low relative sea level, a fault scarp, and a subtle character anomaly in the shape of a fan.

(29) Other important sand bodies of a local type are the elongated sands. These include beach sands, barrier sands, bar-finger sands, abandoned distributary channels, and the meander belts of rivers.

(30) These elongated sands may intercept only a portion of the reflection zone; the amplitude of their seismic reflections is correspondingly reduced—an effect which worsens with increasing depth. The appeal of a gas-sand reservoir remains fairly conveyed by its reflection amplitude, under these circumstances; the appeal of an oil reservoir is falsified.

(31) The association of corresponding onlap indications on many seismic lines allows us to determine the direction of a shoreline. Then, with the environment established, and its trend, we search for anomalies of reflection character in positions corresponding to the above types of elongated sands. Again the key lies in the **mapping** of character or amplitude changes.

(32) Those elongated sands expected in association with a delta are the easiest to locate in a general sense—because of the rather clear-cut seismic indications of a delta—but we expect complications represented by the superimposition of many positions of the delta. Likely directions for the long axis of these bodies can be assessed from the direction of outbuilding of the delta.

(33) Beach and barrier sands, obviously, are aligned with the shoreline, while meander belts are generally at right angles.

(34) A likely location for strike-valley sands can be sought whenever the onlap signature overlies a truncated unconformity.

(35) Turbidites are sought by identifying a deep-water environment and the shelf edge, and by mapping mounded anomalies—particularly those seen to have a possible origin at the mouth of a submarine canyon. The submarine fan at the mouth of the canyon is a prime exploration objective; very extensive and very thick sandstones may be expected in the upper part of the suprafan lobes. Further, the seismic signature of the fan itself may be clearly recognizable and interpretable

from its form, whereas the seismic response to distal turbidite mounds is often weak and obscure.

(36) Of course, all these strategies are idealized. In the presence of a host of complicating factors (subsidence, rotation, folding, faulting and so on) the essential interpretive qualities are not concerned with memorizing an atlas of idealized models, but with geological vision and an understanding of the relation between real geology and its seismic expression.

(37) Item 13 above—to the effect that sandstones cannot form without sand—is a caution to all strategies of prospecting for a sand environment. It is inevitable that we shall always drill more dry holes in this type of work than in prospecting for structure in the traditional way.

(38) The seismic method can now make real contribution to the problem of locating step-out or delineating wells, after a fortuitous discovery. In this case the knowledge obtained from the discovery well can be used to optimize the seismic field work and processing. This is important; we value every improvement which can be made in the seismic indications of a productive anomaly, and in favourable cases these are considerable.

(39) In particular, new and important possibilities are open to us if we use a borehole and a profile of surface source positions.

(40) Every petroleum geologist and geophysicist must read AAPG Memoir 26—Seismic Stratigraphy—from cover to cover.

(41) Twice.

THE END

The final pages of these notes include a number of seismic sections, not subdivided as to the type of targets which they show. Perhaps, to provide us with an informal conclusion to our course, you would care to share with all of us your thoughts on these sections and their potential targets.

But everyone had already left. Thank God it's Friday.

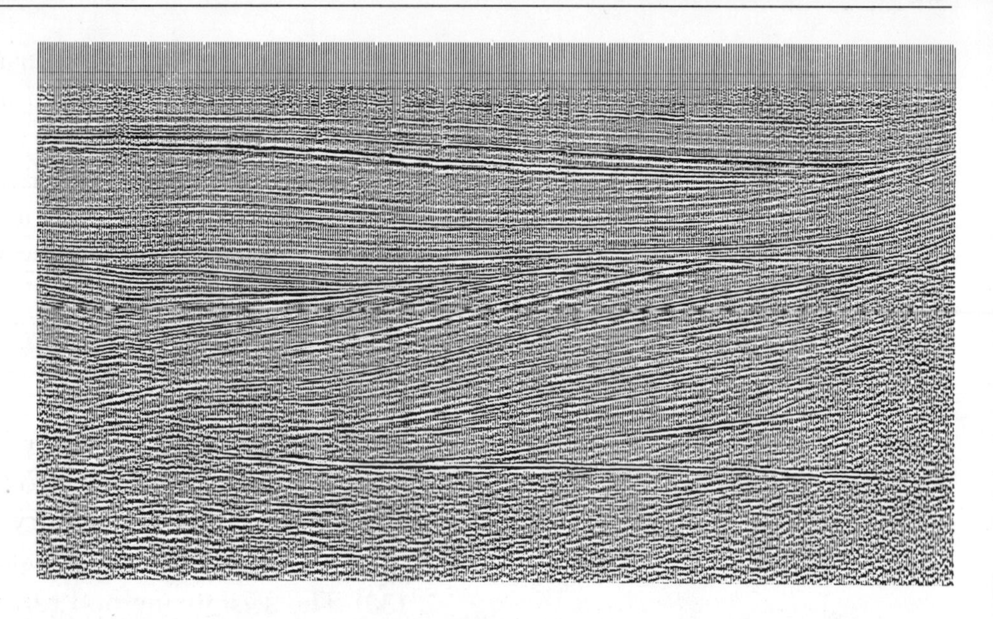

FIGURE 5.1 *(Courtesy Prakla-Seismos)*

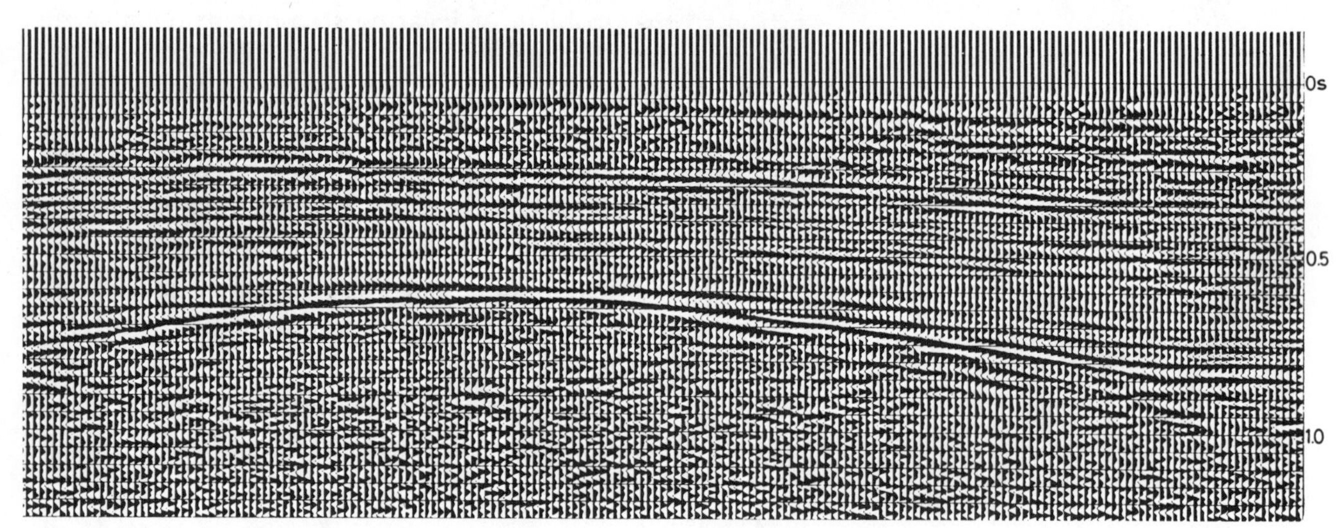

FIGURE 5.2 *(Courtesy Prakla-Seismos)*

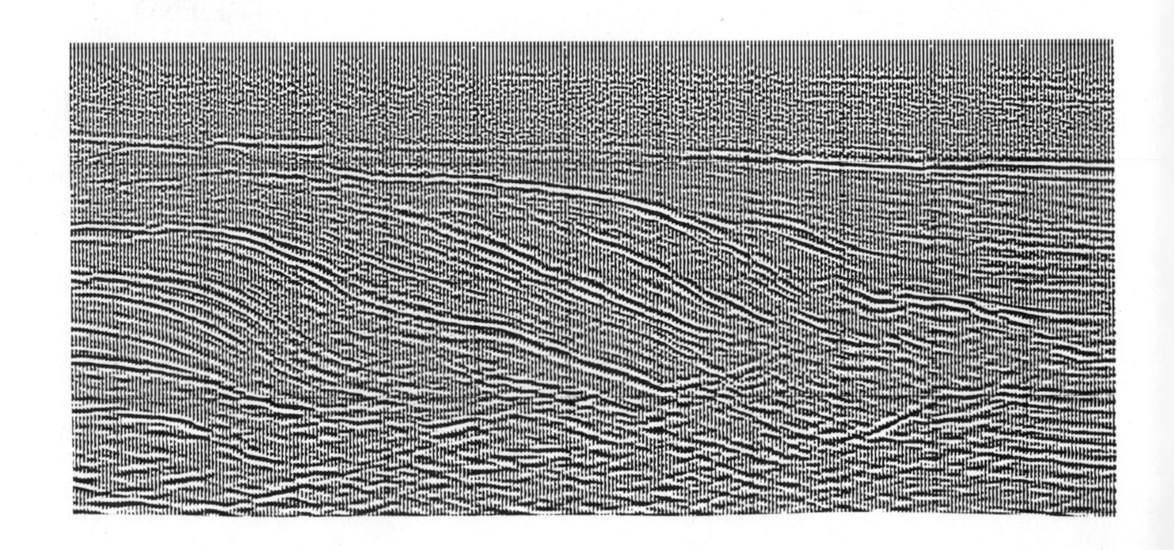

FIGURE 5.3

FIGURE 5.4

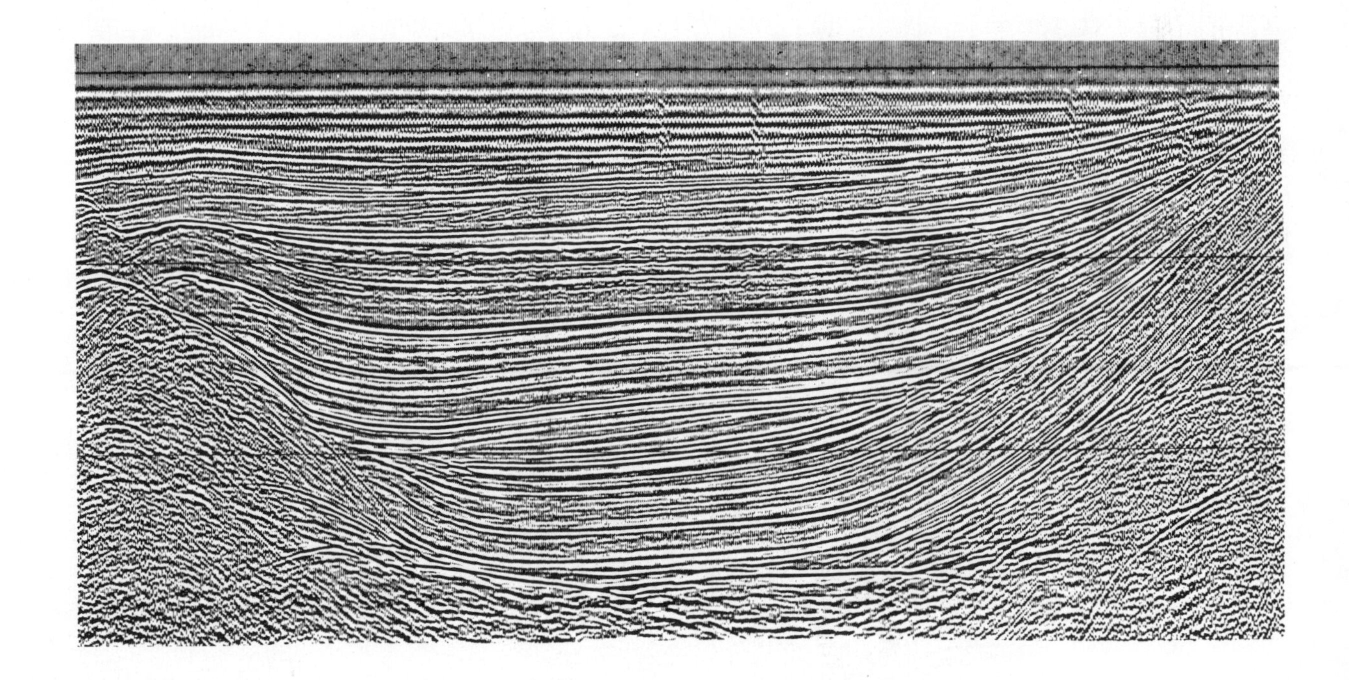

FIGURE 5.5

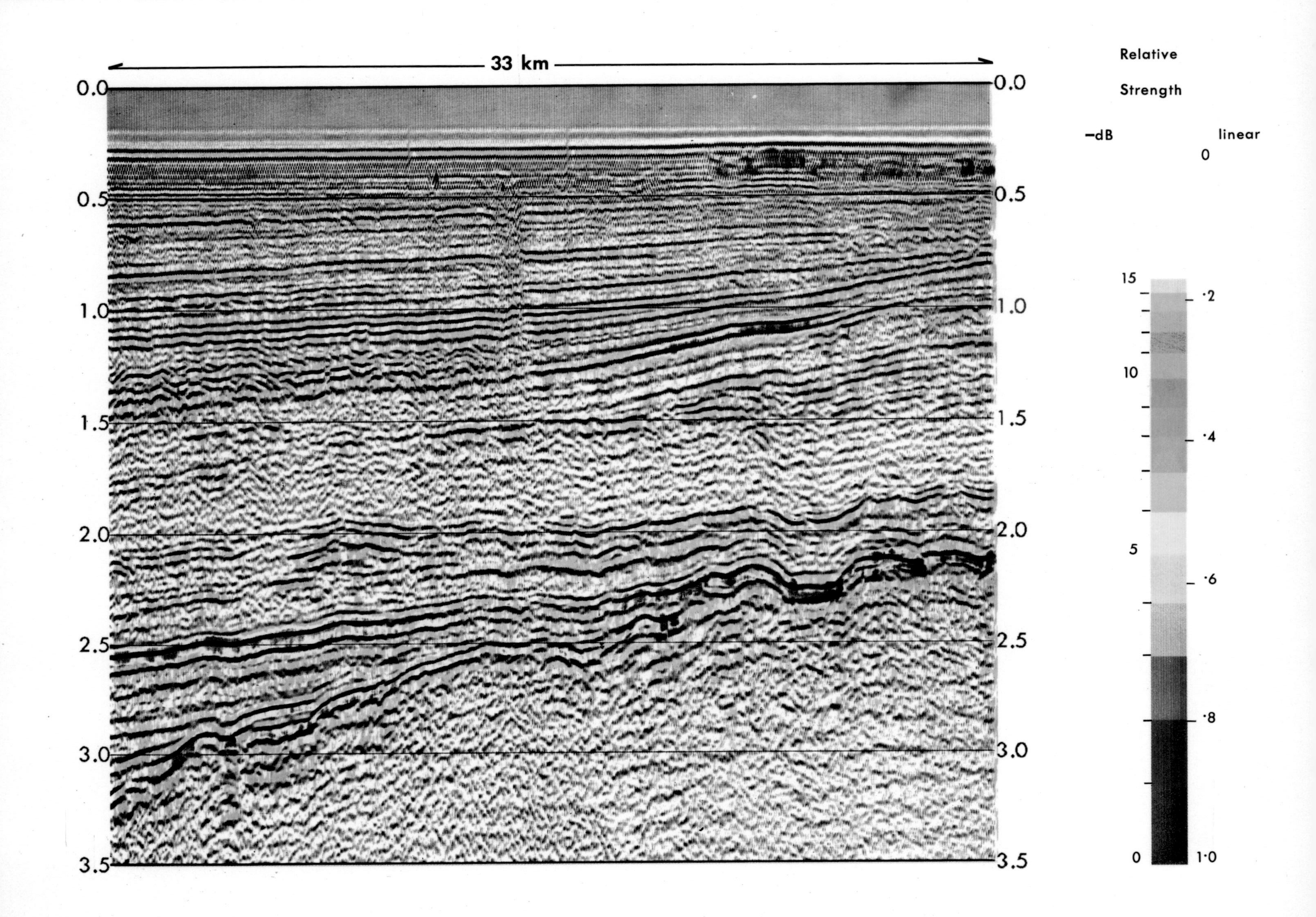

33 km
0.0
0.5
1.0
1.5
2.0
2.5
3.0
3.5
Relative
Strength
-dB
linear
0
15
10
5
0
·2
·4
·6
·8
1·0

References

Becquey, M., Lavergne, M., and Willm, C., 1978. *Acoustic Impedance Logs Computed from Seismic Traces.* Institut Francais du Petrole.

Bruce, C.H., 1973. Pressured Shale and Related Sediment Deformation: Mechanism for Development of Regional Contemporaneous Faults. *Bull. Amer. Assoc. Petrol. Geol.*, **57**, p. 878.

Busch, D.A., 1974. Stratigraphic Traps in Sandstones—Exploration Techniques. *Amer. Assoc. Pet. Geol. Memoir,* No. 21.

—, 1978. Exploration Methods for Sandstone Reservoirs. Course sponsored by Oil and Gas Consultants International, Inc.

Coleman, J.M., 1976. Deltas: Processes of Deposition and Models for Exploration. Continuing Education Publication Co., Champaign, IL. 102 p.

Curry, W.H., and Curry, W.H. III, 1972. South Glenrock Oil Field, Wyoming. *Amer. Assoc. Pet. Geol. Memoir,* No. 16, p. 415.

Eckelmann, W.R., Dewitt, R.J., and Fisher, W.L., 1975. Prediction of Fluvial-Deltaic Reservoir Geometry, Prudhoe Bay Field, Alaska. Proceedings of the 9th World Petroleum Congress. Panel Discussion 4 (1).

Evamy, D.D., Haremboure, J., Kamerling, P., Knapp, W.A., Molloy, F.A., and Rowlands, P.H., 1978. Hydrocarbon Habitat of Tertiary Niger Delta. *Bull. Amer. Assoc. Petrol. Geol.*, **62** p. 1.

Farr, J.B., 1976. How High is High Resolution? Paper presented to the 46th Annual Meeting of the SEG, Houston.

Fisk, H.N., 1961. Bar-finger Sands of the Mississippi Delta. In: Geometry of Sandstone Bodies. *Amer. Assoc. Petrol. Geol. Memoir,* p. 29.

Galloway, W.E., Yancey, M.S., and Whipple, A.P., 1977. Seismic Stratigraphic Model of Depositional Platform Margin, Eastern

Anadarko Basin, Oklahoma. *Amer. Assoc. Pet. Geol. Memoir,* No. 26, p. 439.

Gould, H.R., 1970. The Mississippi Delta Complex. *Soc. Econ. Pal. Min.,* Sp. Pub., No. 25, p. 3.

King, R.E., 1972. Exploration for Stratigraphic Traps—Present Status and Future Outlook. *Amer. Assoc. Pet. Geol. Memoir,* No. 16, p. 643.

Lavergne, M., Willm, C., and Lacaze, J., 1978. *Lithological Determination in Sand-Shale Sequences.* Institut Francais du Petrole (ref. 26456).

Le Blanc, R.J., 1972. Geometry of Sandstone Reservoir Bodies. *Amer. Assoc. Pet. Geol. Memoir,* No. 18, p. 133.

Leung, C.C., and Tree, E.L., 1977. Extended Array Simulation and Application. Paper presented to the 47th Annual Meeting of the SEG, Calgary.

Lindseth, R.O., 1979. Synthetic Seismic Logs—A Process for Stratigraphic Interpretation. *Geophysics,* **44,** No. 1, p. 3.

Lindsey, J.P., Dedman, E.V., and Ausburn, B.E., 1978. New Seismic Techniques Define Strat, Lith Details. *World Oil,* May,1978, p. 56.

Louis, P.R., and Mermey, P.A., 1979. An Example of Seismic Stratigraphy: The Porcupine Basin, Western Ireland. Paper presented to the 49th Annual Meeting of the SEG, New Orleans.

Lyons, P.L., and Dobrin, M.B., 1972. Seismic Exploration for Stratigraphic Traps. *Amer. Assoc. Pet. Geol. Memoir,* No. 16, p. 225.

MacKenzie, D.B., 1972. Primary Stratigraphic Traps in Sand stones. *Amer. Assoc. Pet. Geol. Memoir,* No. 16, p. 47.

McGregor, A.A., and Biggs, C.A., 1972. Bell Creek Oil Field, Montana. *Amer. Assoc. Pet. Geol. Memoir,* No. 16, p. 47.

Payton, C.E. (Ed.), 1977. *Seismic Stratigraphy—Applications To Hydrocarbon Exploration.* AAPG, Tulsa, 503 p.

Roksandic, M.M., 1978. Seismic Facies Analysis Concepts. *Geophysical Prospecting,* **26,** No. 2, p. 383.

Sangree, J.B., and Widmier, J.M., 1979. Interpretation of Depositional Facies from Seismic Data. *Geophysics,* **44,** No. 2, p. 131.

Selley, R.C., 1976. *An Introduction to Sedimentology.* Academic Press, New York, 408 p.

—, 1978. *Ancient Sedimentary Environments.* Cornell University Press, Ithaca, 287 p.

Shelton, J.W., 1967. Stratigraphic Models and General Criteria for Recognition of Alluvial, Barrier Bar and Turbidity-current Sand Deposits. *Bull. Amer. Assoc. Petrol. Geol.,* **51,** No. 12, p. 2441.

Siemers, C.T., 1979. Submarine-Fan Deposition of Woodbine-Eagle Ford Interval. *Oil & Gas Journal,* September 3, 10, 17, 1979.

Tufekcic, D., 1978. A Prediction of Sedimentary Environment from Marine Seismic Data. *Geophysical Prospecting,* **26,** No. 2, p. 329.

Walker, R.C., 1978. Deep-water Sandstone Facies and Ancient Submarine Fans. *Bull. Amer. Assoc. Petrol. Geol.,* **62,** No. 6, p. 932.

Weber, K.J., and Daukoru, E., 1975. Petroleum Geology of the Niger Delta. Proceedings of the 9th World Petroleum Congress, Panel Discussion 4 (1).

Widess, M.B., 1954. How Thin is a Thin Bed? *Geophysics,* **38,** No. 6, p. 1176.